Laboratory Ma

M000248741

to accompany

Integrated Science

Fifth Edition

Bill W. Tillery
Arizona State University

Eldon D. Enger
Delta College

Frederick C. Ross
Delta College

Laboratory Manual to accompany
INTEGRATED SCIENCE, FIFTH EDITION
BILL W. TILLERY, ELDON D. ENGER, AND FREDERICK C. ROSS

Published by McGraw-Hill Higher Education, an imprint of The McGraw-Hill Companies, Inc., 1221 Avenue of the Americas, New York, NY 10020. Copyright © 2011, 2008, 2007, and 2001 by The McGraw-Hill Companies, Inc. All rights reserved.

No part of this publication may be reproduced or distributed in any form or by any means, or stored in a database or retrieval system, without the prior written consent of The McGraw-Hill Companies, Inc., including, but not limited to, network or other electronic storage or transmission, or broadcast for distance learning.

1 2 3 4 5 6 7 8 9 0 WDQ/WDQ 0

ISBN: 978-0-07-729286-7
MHID: 0-07-729286-3

Contents

Appendix ....**Page**

Introduction

The "big picture" of laboratory work is an opportunity to "get your hands on science" beyond reading and actually enter into the process of doing science. The benefit you derive from the laboratory is directly dependent on two things: first, the care with which you perform the experiments and record your observations, and second, your awareness of the relationship between your observations and the general principles under study.

In more detail, laboratory work provides opportunities for you to learn the practical knowledge necessary for a well-rounded understanding of the sciences. You will learn the theory and scientific laws pertaining to natural phenomena in the lecture portion of the course. Conducting experiments and collecting data to test the validity of our understandings requires a different set of skills than those required for success in the lecture. Success in the laboratory involves skills in making accurate measurements in the real world, then formulating valid generalizations and principles based on the data. Every experiment in the laboratory will provide lessons and opportunities to learn these skills. This laboratory manual includes open-ended *Invitations to Inquiry* that can be used as stand-alone investigations or as special projects along with the more structured learning environment. In addition, there is a *Special Project* (Lab 44) designed to serve as a guide for a longer term, open-ended student investigation.

The instructions for each experiment in a structured learning environment include some basic theory and relationships about the phenomena to be investigated, data tables, graph paper, and questions to be answered. These are not "fill in the blank" type questions as are typically found in a laboratory workbook but are designed to help students make a thoughtful analysis with careful, thorough thinking. The instructions for inquiry experiments, on the other hand, are more toward suggestions for interesting, challenging investigations. Less student support is provided for the inquiry experiments, which are usually more successfully completed by more mature students.

For the student. Always bring this laboratory guide to each session, completing required observations and calculations as you work. Be as accurate and neat as possible, but do not waste time on a report. Sloppy work is to be avoided, of course, but *concentrate on knowing what you are doing, gathering accurate data, and working out conclusions* while in the laboratory. These are the criteria for evaluating laboratory reports, not how much time you spend detailing a handsome report. Focus on these criteria during each laboratory session. In physics experiments, this means accomplishing the purpose within a reasonable error, reasoning what constitutes a reasonable error, and stating what the origin of this error might be. In most of the labs, the purpose is to verify a law or theory that you have covered in class. Therefore, the conclusion may be as simple as "Newton's third law of motion was verified within 10% of the expected value." This should be followed by a statement as to why 10% is an acceptable error for this particular experiment. The conclusions should be reasonable and make sense, not necessarily agreeing with "expected" findings. Thoughtful analysis and careful, thorough thinking are much more important (and reasonable) than 0% error. In chemistry experiments, the purpose of a lab is often to understand how matter interacts and the procedures and tests used in understanding matter. The results of procedures and tests are often used to analyze and reason and to identify unknowns. Analysis of findings and possible errors are appropriate in these laboratory experiments as well as in the biological science experiments. In general, you will have the opportunity to measure, experiment, observe, and discover for yourself in all the integrated sciences.

You are encouraged to work together in study groups, but your work should be your own. Note all the appendices at the end of this guide. They cover most of the problems that past students have had with laboratory skills and procedures.

Acknowledgments

The astronomy investigations of Celestial Coordinates, Motions of the Sun, Phases of the Moon, and Power Output of the Sun are from the astronomy manual of *Laboratory Astronomy: Experiments and Exercises*, Anthony J. Nicastro, Wm. C. Brown Publishers, 1990.

The biology investigations are from *Laboratory Manual: Concepts in Biology,* 10th Edition, Eldon D. Enger and Frederick C. Ross, McGraw-Hill Publishers, 2003.

The alternative materials were suggested by Franz Fry, Concord University.

Materials Required for Each Experiment
(Quantities given are for an individual or teams of students.)

1. **Graphing** (page 1): Meterstick, masking tape, several types of small balls that bounce (tennis, ping-pong, rubber hand ball, etc.).

2. **Ratios** (page 15): Several sizes of cups or beakers with round bottoms for tracing circles, metric rulers, string, wood or plastic cubes (about 30), three sizes of rectangular containers to hold water, balance, graduated cylinder.

3. **Motion** (page 27): Battery-operated toy bulldozer (or other toy car); long sheets of computer paper, butcher paper, or adding machine paper; masking tape; meterstick; stopwatch; inclined ramp (1 m or longer); 1 to 2 m rolling ball track; steel ball or marble (alternate setup: lab cart, photogates, computer with timer software).

4. **Free Fall** (page 43): Free-fall apparatus with spark timer, mass, meterstick, metric ruler (alternate setup: lab cart on track, photogate and pulley with spokes, computer with timer software).

5. **Centripetal Force** (page 53): Mass hanger and small masses, nylon string, rubber stopper, holder (wood or plastic rod with small hole for string; see figure 5.1).

6. **Work and Power** (page 59): Stairwell, meterstick, stopwatch, a scale for weighing people.

7. **Thermometer Fixed Points** (page 63): Laboratory thermometer, beakers, crushed ice, steam generator, laboratory mercury barometer.

8. **Specific Heat** (page 71): Two Styrofoam cups to serve as a calorimeter, balance, three samples of shot made of different metals (e.g., aluminum, copper, lead), heating source for boiling water (hot plate and beaker or burner and ring stand setup), thermometer.

9. **Speed of Sound in Air** (page 79): Resonance tube apparatus, meterstick, two tuning forks of different pitches, rubber hammer, thermometer.

10. **Static Electricity** (page 87): Two glass rods, two hard rubber rods, nylon or silk cloth, wool cloth or fur, thread, electroscope, two rubber balloons.

11. **Ohm's Law** (page 93): Adjustable dc power supply, dc voltmeter, dc ammeter, resistors, hookup wire or patch cords.

12. **Magnetic Fields** (page 103): Large sheet of unlined white paper, small magnetic compass, bar magnet, sharp pencil, large plastic sheet or glass plate, iron filings.

13. **Reflection and Refraction** (page 107): Ruler, cardboard (from box), small flat mirror, small wood block, rubber bands, straight pins, unlined white paper, protractor, 5 cm square glass plate.

14. **Physical and Chemical Change** (page 113): Sodium chloride, graduated cylinder, evaporating dish, nichrome wire, tongs, magnesium ribbon, small test tube, dilute hydrochloric acid, silver nitrate solution, funnel, funnel support, ring stand, filter paper, copper(II) chloride solution, beaker, aluminum foil. **Alternative Suggestion:** Use calcium chloride and sodium carbonate or sodium oxalate to form a precipitate instead of silver nitrate and hydrachloric acid.

15. **Conductivity of Solutions** (page 119): Conductivity apparatus (figure 15.1), short patch cord with alligator clips, distilled water, tap water, dry sodium chloride (table salt), sugar solution (1%), ethyl alcohol, hydrochloric acid (1.0 M: 81 mL concentrated/L solution), sodium hydroxide (1.0 M: 40 g solid/L solution), sodium chloride solution (1.0 M: 58 g solid/L solution), vinegar, glycerine.

16. **Metal Replacement Reactions** (page 125): Metal strips (about 2 cm × 5 cm) of copper, zinc, and lead; sandpaper; test tubes; test tube rack; graduated cylinder; 10 cm length of thin copper wire; silver nitrate solution (0.1 M: 17.0 g/L of solution); copper nitrate solution (0.1 M: 24.2 g/L of solution); lead nitrate solution (0.1 M: 33.1 g/L of solution). **Alternative Suggestion:** The four metals involved can be copper solution, tin and tin solution, zinc and zinc solution, and aluminum. This list will still give the same metal replacement based on activity and remove the silver and lead.

17. **Identifying Salts** (page 131): Beakers, burner, medicine dropper, cobalt glass squares, platinum or nichrome wire, forceps, test tubes and rack, graduated cylinder, pipette (or thin glass tube). Flame test solutions: sodium nitrate solution (1.0 M: 85 g/L of solution), lithium nitrate (1.0 M: 123 g/L of solution), strontium nitrate (1.0 M: 284 g/L of solution), calcium nitrate (1.0 M 163 g/L of solution), barium nitrate (1.0 M: 261 g/L of solution), potassium nitrate (1.0 M: 101 g/L of solution), dilute hydrochloric acid (1:4), and distilled water. Chemical test solutions: silver nitrate solution (0.1 M: 17.3 g/L of solution), barium chloride solution (0.1 M: 24.4 g/L of solution), calcium carbonate solution (saturated), iron(II) sulfate solution (saturated), concentrated sulfuric acid, sodium chloride solution (0.1 M: 5.9 g/L of solution), potassium sulfate solution (0.1 M: 15.8 g/L of solution). **Alternative Suggestion:** Boron produces a nice green flame in a flame test, and boric acid can replace barium nitrate. Calcium nitrate can also replace barium nitrate for the sulfate test. Silver nitrate is still the best ion for the chloride test and the other alternatives are also regulated or unstable. Positive results should be centrifuged to remove the solid silver chloride. The liquid portion should be collected in a separate container. Sodium chloride can then be added and any solid should be removed. The liquid can then go down the drain.

18. **Natural Water** (page 137): Distilled water, 3% alum solution, aqueous ammonium hydroxide (6.0 M), beakers, dropper, glass stirring rod, filtering funnel, ring stand and ring, glass wool, clean sand, test tubes and rack, wire screen, burner, watch glass, graduated cylinder, calcium chloride solution (0.001 M: 0.1 g/L of solution), standard soap solution, muddy water, tap or well water.

19. **Measurement of pH** (page 145): Solutions of hydrochloric acid (0.1 M: 8.1 mL concentrated/L of solution), sulfuric acid (0.1 M: 5.6 mL concentrated/L of solution), acetic acid (0.1 M: 5.6 mL concentrated/L of solution), sodium hydroxide (0.1 m: 4 g/L of solution), barium hydroxide (17 g/L of solution), household ammonia (diluted with 6 volumes water). Indicators: red litmus paper, blue litmus paper, bromthymol blue, methyl orange, methyl red, phenolphthalein, universal indicator paper (with color-coded pH scale). Watch glass, glass stirring rod, test tubes and rack. Various unknown solutions (different acid or base solutions of different concentrations).

20. **Amount of Water Vapor in the Air** (page 153): Sling psychrometer (or two thermometers with a 3 cm length of cotton shoelace on the bulb end of one), meterstick.

21. **Growing Crystals** (page 159): Large Pyrex beaker (1 liter or larger), selected salt or salts (see Table 21.1 for possibilities and approximate amounts of salt required for each), ring stand and ring, wire screen, burner, balance, glass stirring rod, glass jar (1 liter or larger with lid), spatula, very fine fishing leader, tape, epoxy glue. **Alternative Suggestion:** Potassium chromate should not be used.

22. **Properties of Common Minerals** (page 165): Mineral collection, streak plate, magnifying glass, steel file, pocketknife, copper penny, magnet, dilute hydrochloric acid, dropper, balance, graduated cylinder, overflow can to determine density (optional).

23. **Density of Igneous Rocks** (page 171): Balance, overflow can, graduated cylinder, ring stand and ring, thin nylon string, beaker, granite specimen, basalt specimen.

24. **Latitude and Longitude** (page 175): Soft clay (fist-sized lump), protractor, toothpicks, knife, pencil.

25. **Telescopes** (page 181): Convex lenses (long and short focal length), lens holders, meterstick, meterstick supports, tagboard screen, holder for tagboard screen, luminous object (clear glass light bulb or candle).

26. **Celestial Coordinates** (page 189): Celestial globe, self-stick ("sticky") notes.

27. **Motions of the Sun** (page 199): None required.

28. **Diffusion and Osmosis** (page 209) For 10 laboratory groups: 60 4-inch pieces of dialysis tubing, a ball of string, 4 L of molasses, balance, 10 funnels, 30 400-mL beakers, 10 ring stands/rings, screens, Bunsen burners (or hot plates or water baths), 10 thermometers, crushed ice, 10 10-mL graduated cylinders, top loading balance or other appropriate balances.

29. **The Microscope** (page 223): 25 compound microscopes, lens tissues, supply of slides, supply of coverslips, beakers of water/droppers, various types of prepared slides, human hair, variety of materials such as protozoa, cork, potato, algae, or microscopic animal life.

30. **Survey of Cell Types: Structure and Function** (page 235): Compound microscopes (one for each student is desirable, but it is possible to accomplish the objectives with fewer); blank slides; coverslips; sharp, single-edge razor blades; 5 dropper bottles of deionized water; 5 dropper bottles of methyl cellulose or protoslo; 5 dropper bottles of Lugol's solution; 5 dropper bottles of methylene

blue solution; 5 dropper bottles of 5% or 10% salt solution; fresh Elodea; fresh onion (one onion is more than enough for a lab of 25 students); cultures of the following: Anabaena, Spyrogyra, Euglena, Paramecium; culture of soil bacteria (A culture of soil bacteria can be produced by introducing some soil into nutrient broth a few days prior to the lab. There will be a variety of bacteria present, and many of them will be moving which makes it easy for students to pick out.); culture of mold; 6 forceps; 1 box toothpicks; yeast cells stained with the red dye; congo red; 5 beakers with antibiotic soap for students to use for toothpick and cheek epithelial slide disposal.

31. **Enzymes** (page 247) For 6 laboratory groups: 6 large trays with the following in each tray: 1 test tube rack, 21 test tubes, 1 small dropper bottle of pyrocatechol (labeled substrate—0.66 g pyrocatechol in 100 mL water; adjust pH to between 6.0 and 6.5. One week shelf life under refrigeration.), 2 250-mL beakers, 1 thermometer, 1 small dropper bottle of enzyme (labeled potato juice—0.05 g tyrosinase in 100 mL water. One week shelf life under refrigeration.), 1 small dropper bottle of phenylthiourea (0.001 g in 100-mL water. Long shelf life at room temperature.), 1 small dropper bottle of tyrosine (0.05 g tyrosinase in 100 mL water 0.05% solution. Long shelf life at room temperature.), 1 small dropper bottle of 5% sucrose solution, glass marking pencil, water baths at 0°, 20°, 40°, 60°, 100° C, distilled water, masking tape, crushed ice, pHydrion paper or pH meter, 100 mL water at pH 3, 100 mL water at pH 5, 100 mL water at pH 7, 100 mL water at pH 9, 100 mL water at pH 9, 100 mL water at pH 11.

32. **Photosynthesis and Respiration** (page 259) For 6 laboratory groups: 6 large trays with the following on each tray: 2 test tube racks that will take 25 × 200 mm test tubes, 8 25 × 200-mm test tubes with corks, 1 400-mL beaker, 1 Hatch kit for dissolved oxygen test, 1 Hatch kit for dissolved carbon dioxide test, supply of healthy *Elodea* or other aquatic plants, 12 goldfish or other small fish, small fish net, 6 individual fluorescent lights (or bank of fluorescent lights to be used by all groups), 3 gallons aged stock water aged at least 48 hours (bubbling air through the water is a good idea, since the water will be saturated with oxygen), cabinet or box to provide dark environment.

33. **The Chemistry and Ecology of Yogurt Production** (page 267): 25 ½-pint containers of milk, hot plates, 25 400-mL glass beakers, 275 g dehydrated milk powder (11 g per student), 25 thermometers, 25 stirring rods, stapler, masking tape, various flavoring materials (optional), 1 package of freeze-dried yogurt culture containing *Streptococcus thermophilus* and *Lactobacillus bulgaricus*.

34. **DNA and RNA** (page 273): 24 DNA kits for student use. Combine the contents from two stock DNA kits to make one kit with the number of parts listed in the procedure portion of the lab exercise. There will be extra parts left over for use as spares. Relabel the parts with appropriate letter designations; A = adenine, P = phosphate, D = deoxyribose, etc. Place appropriate names on the three amino acids.

35. **Mitosis—Cell Division** (page 289): 25 microscopes, 25 Allium root tip slides, 25 whitefish blastula (or other suitable slide).

36. **Meiosis** (page 299) For 12 laboratory groups: 12 sets of chromosome models for a class of 24. Each set consists of 4 chromosomes. See diagram in exercise. Construct models of chromosomes using pop beads, magnets, and plastic sleeves with slips of paper inside. 12 large sheets of newsprint or white boards. Materials needed to make 15 sets of chromosome models are as follows: 900 red

pop beads, 900 yellow pop beads (available from Carolina Biological, Burlington, NC 27215, Catalogue #17 1112), 120 magnetic centromeres (available from Carolina Biological, Burlington, NC 27215, Catalogue #17 1114), 50 feet of clear plastic tubing.

37. **Genetics Problems** (page 311): A copy of the laboratory exercise for each student.

38. **Human Variation** (page 327) For 24 students: 12 dry marker boards; 30 dry markers, black, brown, tan, yellow, red, green, blue; 24 coins; 15 erasers; 15 calculators (optional).

39. **Sensory Abilities** (page 343): 1 box of cotton swabs; 100 mL of the following solutions: 0.5% acetic acid, 0.5% quinine, 5.0% NaCl, 5.0% sucrose; NaCl crystals; 13 soft lead pencils; 13 blunt probes; 13 calipers (Forceps may be used if plasticine is wedged in the forceps to keep the points the required distance apart.); 13 nails in ice water; 13 nails in moderately hot water; 13 sets of differently colored paper squares 5 cm × 5 cm.

40. **Daily Energy Balance** (page 355): No equipment is required; however, students will be required to use a calorie guide. These might be obtained in grocery, drug, or variety stores or in the library. An additional valuable aid would be a copy of the latest edition of the Recommended Daily Allowances.

41. **The Effect of Abiotic Factors on Habitat Preference** (page 363): Materials for each of five items include the following: 1 meter length of ½-inch clear plastic tube, 1 meterstick, 4 clamps, 2 corks, 5 100-mL beakers, pH paper, 1 thermometer, 2 petri dishes, 100 mL Ringer's solution (optional), 1 funnel, 150 mL solution of brine shrimp, masking tape, 1 graduated cylinder, overhead projector (optional). Special equipment for various teams: Team 1 control: plastic tube wrapped with opaque tape. Team 2 pH: plastic tube wrapped with opaque tape, 0.5 mL 1% HCl, 1.0 mL 1% KOH, 2 hypodermic needles and syringes. Team 3 temperature: plastic tube wrapped with opaque tape, plastic bag of crushed ice, infrared heat lamp. Team 4 light: Clear plastic tube—only use if the tube is placed in a light tight place with a directional source of light. Alternatively, use 2 tubes wrapped with black tape, to be light tight; 3 tubes with opaque tape; and 3 tubes left clear. Team 5 gravity: plastic tube wrapped with black tape, ring stand to hold tube upright.

42. **Natural Selection** (page 371): Pack of playing cards, Ptc paper, 4 × 8 piece of paper with a word printed on it such as windmill, 10 meter sticks.

43. **Roll Call of the Animals** (page 379): Representatives of the animals listed.

44. **Special Project** (page 383): Supplied by student.

Name_____Section_____Date_____

Experiment 1: Graphing

Invitation to Inquiry

The measurement of a quantity that can have different values at different times is called a **variable**. For example, the rate of your heartbeat, the number of times you breathe per minute, and your blood pressure are all variables because they can have different values at different times. In many situations, there are relationships that occur between variables. The rate of your breathing, for example, increases when you begin to exercise, so you could say that the breathing rate is in **direct proportion** to exercise up to a certain limit.

Measurements of variables that increase or decrease relative to each other are in direct proportion and will yield a straight line on a graph. This relationship is said to be direct, or *linear*. There are more types of relationships between variables, and most can be identified as producing one of five basic shapes of graphs. These are identified, left to right, as no relationship, linear, inverse, square, and square root.

After giving the possibilities some thought, look for relationships that might result in (1) a direct relationship, then (2) something other than a direct relationship. Make measurements, graph your data, then decide which of the five shapes the graph resembles. For example, compare your heartbeat rate before climbing any stairs, then after climbing 10 stairs, 20 stairs, and 30 stairs. What is the shape of a graph comparing the heartbeat rate and the number of stairs climbed? What does this mean about the relationship between the number of stairs climbed and your heartbeat?

What other relationships can you find in the lab, outside, or between any two variables in everyday occurrences? Summarize your findings here:

Background

Refer to figure 1.1 for terminology used when discussing a graph and see Appendix I on page 387 for a detailed discussion about the terms.

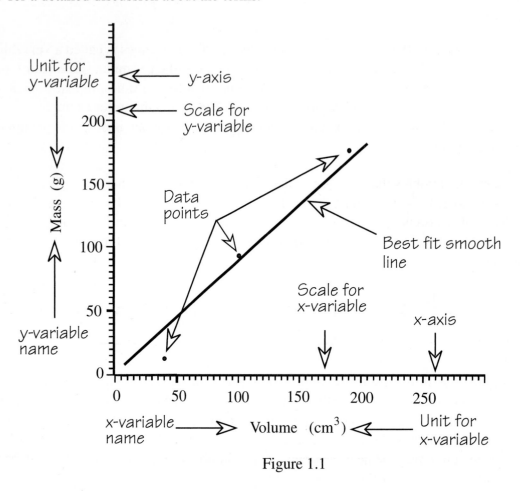

Figure 1.1

Procedure

1. Position a meterstick vertically on a flat surface, such as a wall or the side of a lab bench. Be sure the metric scale of the meterstick is on the outside and secure the meterstick to the wall or lab bench with two strips of masking tape.

2. Drop a ball as close as possible to the meterstick and measure (a) the height dropped and (b) the resulting height bounced. Repeat this for three different heights dropped and record all data in Data Table 1.1 on page 6. In the data table, identify the *independent* (manipulated) variable and the *dependent* (responding) variable.

3. Use the graph paper on page 9 to make a graph of the data in Data Table 1.1, being sure to follow all the rules of graphing (see Appendix I on page 387 for help). Title the graph, "Single Measurement Bounce Height as a Function of Height Dropped."

4. After constructing the graph, but before continuing with this laboratory investigation, answer the following questions:

    (a) What decisions did you have to make about how you conducted the ball-dropping investigation?

    (b) Would you obtain the exact same result if you dropped the ball from the same height several times? Explain.

    (c) Did you make a dot-to-dot line connecting the data points on your graph? Why or why not?

    (d) Could you use your graph to obtain a predictable result for dropping the ball from different heights? Explain why you could or could not.

(e) What is the significance of the origin on the graph of this data? Did you use the origin as a data point? Why or why not?

5. Make at least three more measurements for each of the previous three *height-dropped* levels. Find the *average height bounced* for each level and record the data and the average values in Data Table 1.2 on page 6.

6. Make a new graph of the *average height bounced* for each level that the ball was dropped. Draw a *straight best fit line* that *includes the origin* by considering the general trend of the data points. Draw the straight line as close as possible to as many data points as you can. Try to have about the same number of data points on both sides of the straight line. Title this graph, "Averaged Bounce Height."

7. Compare how well both graphs, "Single Measurement Bounce Height" and "Averaged Bounce Height," predict the heights that the ball will bounce for *heights dropped* that were not tried previously. Locate an untried height-dropped distance on the straight line, then use the corresponding value on the scale for height bounced as a prediction. Test predictions by noting several different heights, then measuring the actual heights bounced. Record your predictions and the actual experimental results in Data Table 1.3 on page 7.

8. Use a new, *different kind of ball* and investigate the bounce of this different ball. Record your single-measurement data for this different ball in Data Table 1.4. Record the averaged data for the height of the bounce for the three levels of dropping in Data Table 1.5. Repeat procedure step 7 for the different kind of ball. Record your predictions and the actual results in Data Table 1.6 on page 8.

9. Graph the results of the different kind of ball investigations onto the two previous graphs. Be sure to distinguish between sets of data points and lines by using different kinds of marks. Explain the meaning of the different marks in a *key* on the graph.

10. What does the steepness (slope) of the lines tell you about the bounce of the different balls?

Results

1. Describe the possible sources of error in this experiment.

2. Describe at least one way that data concerning two variables is modified to reduce errors in order to show general trends or patterns.

3. How is a graph modified to show the best approximation of theoretical, error-free relationships between two variables?

4. Compare the usefulness of a graph showing (a) exact, precise data points connected dot-to-dot and (b) an approximated straight line that has about the same number of data points on both sides of the line.

5. Was the purpose of this lab accomplished? Why or why not? (Your answer to this question should be reasonable and make sense, showing thoughtful analysis and careful, thorough thinking.)

Data Table 1.1	Single Measurement Data: 1st Ball	
Trial	Height Dropped _____variable	Height Bounced _____variable
1	_____	_____
2	_____	_____
3	_____	_____

Data Table 1.2	Averaged Bounce Data: 1st Ball			
	Bounce Height			
Dropped Height	Trial 1	Trial 2	Trial 3	Average
_____	_____	_____	_____	_____
_____	_____	_____	_____	_____
_____	_____	_____	_____	_____

Data Table 1.3   Predictions and Results: 1st Ball

Trial	Single Measurement Data			Averaged Data		
	Dropped Height	Predicted Height	Measured Height	Dropped Height	Predicted Height	Measured Height
1	_____	_____	_____	_____	_____	_____
2	_____	_____	_____	_____	_____	_____
3	_____	_____	_____	_____	_____	_____

Data Table 1.4   Single Measurement Data: 2nd Ball

Trial	Height Dropped	Height Bounced
	_____variable	_____variable
1	_____	_____
2	_____	_____
3	_____	_____

Data Table 1.5	Averaged Bounce Data: 2nd Ball			
	Bounce Height			
Dropped Height	Trial 1	Trial 2	Trial 3	Average
_____	_____	_____	_____	_____
_____	_____	_____	_____	_____
_____	_____	_____	_____	_____

Data Table 1.6	Predictions and Results: 2nd Ball					
Trial	Single Measurement Data			Averaged Data		
	Dropped Height	Predicted Height	Measured Height	Dropped Height	Predicted Height	Measured Height
1	_____	_____	_____	_____	_____	_____
2	_____	_____	_____	_____	_____	_____
3	_____	_____	_____	_____	_____	_____

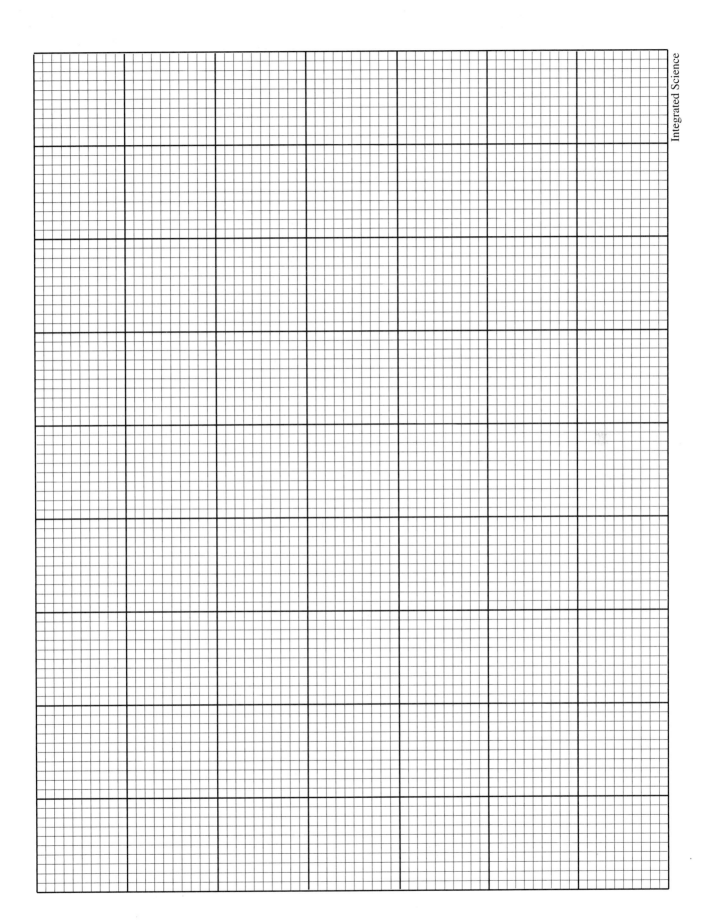

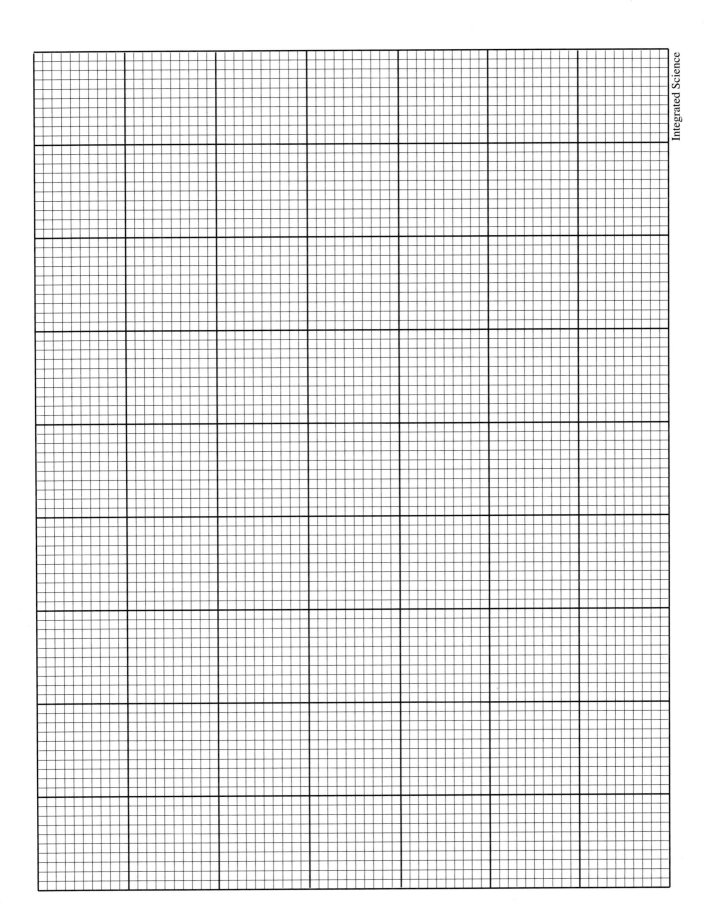

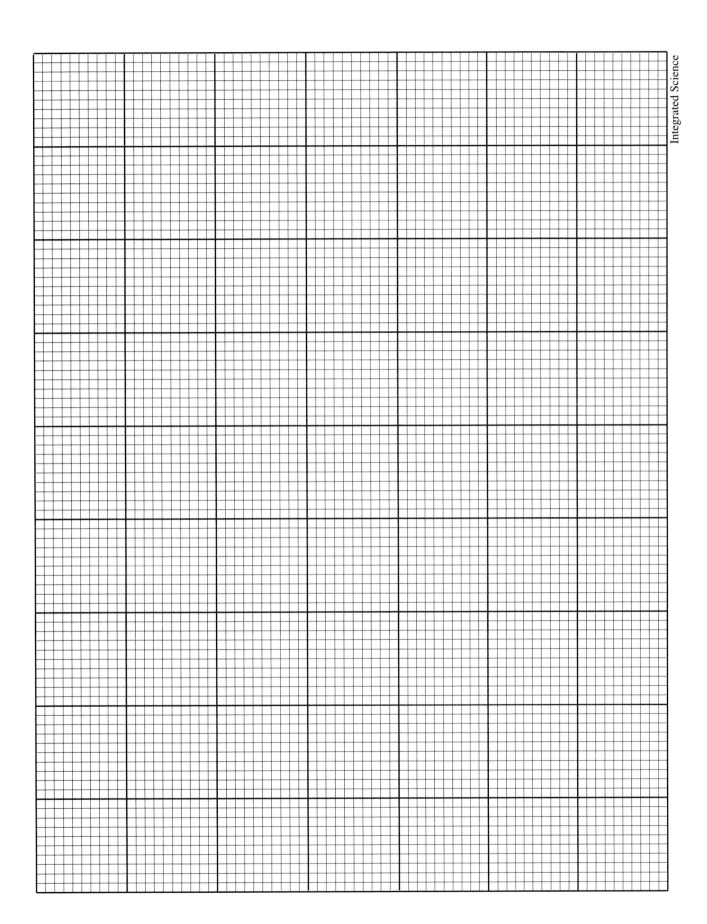

Experiment 2: Ratios

Invitation to Inquiry

If you have popped a batch of popcorn, you know that a given batch of kernels might pop into big and fluffy popcorn. But another batch might not be big and fluffy and some of the kernels might not pop. Popcorn pops because each kernel contains moisture that vaporizes into steam, expanding rapidly and causing the kernel to explode or pop. Here are some questions you might want to consider investigating to find out more about popcorn: Does the ratio of water to kernel mass influence the final fluffy size of popped corn? (Hint: measure mass of kernel before and after popping). Is there an optimum ratio of water to kernel mass for making bigger popped kernels? Is the size of the popped kernels influenced by how rapidly or how slowly you heat the kernels? Can you influence the size of popped kernels by drying or adding moisture to the unpopped kernels? Is a different ratio of moisture to kernel mass better for use in a microwave than in a conventional corn popper? Perhaps you can think of more questions about popcorn.

Summarize your findings here:

Figure 2.1

Background

The purpose of this introductory laboratory exercise is to investigate how measurement data are simplified in order to generalize and identify trends in the data. Data concerning two quantities will be compared as a **ratio**, which is generally defined as a relationship between numbers or quantities. A ratio is usually simplified by dividing one number by another.

Procedure

Part A: Circles and Proportionality Constants

1. Obtain three different sizes of cups, containers, or beakers with circular bases. Trace around the bottoms to make three large but different-sized circles on a blank sheet of paper.

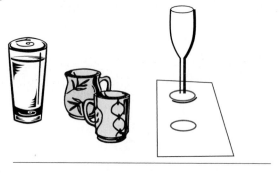

Figure 2.2

2. Mark the diameter on each circle by drawing a straight line across the center. Measure each diameter in millimeters (mm) and record the measurements in Data Table 2.1. Repeat this procedure for each circle for a total of three trials.

3. Measure the circumference of each object by carefully positioning a length of string around the object's base, then grasping the place where the string ends meet. Measure the length in millimeters and record the measurements for each circle in Data Table 2.1. Repeat the procedure for each circle for a total of three trials. Find the ratio of the circumference of each circle to its diameter. Record the ratio for each trial in Data Table 2.1 on page 23.

4. The ratio of the circumference of a circle to its diameter is known as **pi** (symbol π), which has a value of 3.14... (the periods mean many decimal places). Average all the values of π in Data Table 2.1 and calculate the experimental error.

16

Part B: Area and Volume Ratios

1. Obtain one cube from the supply of same-sized cubes in the laboratory. Note that a cube has six sides or six units of surface area. The side of a cube is also called a *face*, so each cube has six identical faces with the same area. The overall surface area of a cube can be found by measuring the length and width of one face (which should have the same value) and then multiplying (length)(width)(number of faces). Use a metric ruler to measure the cube, then calculate the overall surface area and record your finding for this small cube in Data Table 2.2 on page 23.

2. The volume of a cube can be found by multiplying the (length)(width)(height). Measure and calculate the volume of the cube and record your finding for this small cube in Data Table 2.2.

3. Calculate the ratio of surface area to volume and record it in Data Table 2.2.

4. Build a medium-sized cube from eight of the small cubes stacked into one solid cube. Find and record (a) the overall surface area, (b) the volume, and (c) the overall surface area to volume ratio and record them in Data Table 2.2.

5. Build a large cube from 27 of the small cubes stacked into one solid cube. Again, find and record the overall surface area, volume, and overall surface area to volume ratio and record your findings in Data Table 2.2.

6. Describe a pattern, or generalization, concerning the volume of a cube and its surface area to volume ratio. For example, as the volume of a cube increases, what happens to the surface area to volume ratio? How do these two quantities change together for larger and larger cubes?

Part C: Mass and Volume

1. Obtain at least three straight-sided, rectangular containers. Measure the length, width, and height *inside* the container (you do not want the container material included in the volume). Record these measurements in Data Table 2.3 (page 24) in rows 1, 2, and 3. Calculate and record the volume of each container in row 4 of the data table.

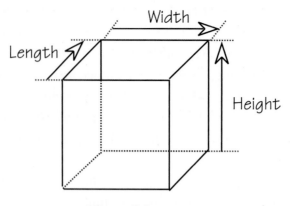

Width

Length

Height

Figure 2.3

2. Measure and record the mass of each container in row 5 of the data table. Measure and record the mass of each container when "level full" of tap water. Record each mass in row 6 of the data table. Calculate and record the mass of the water in each container (mass of container plus water minus mass of empty container, or row 6 minus row 5 for each container). Record the mass of the water in row 7 of the data table.

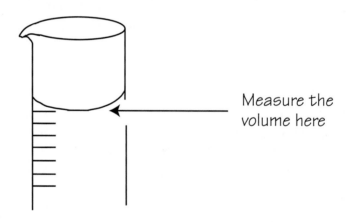

Measure the volume here

Figure 2.4

3. Use a graduated cylinder to measure the volume of water in each of the three containers. Be sure to get *all* the water into the graduated cylinder. Record the water volume of each container in milliliters (mL) in row 8 of the data table.

4. Calculate the ratio of cubic centimeters (cm^3) to mL for each container by dividing the volume in cubic centimeters (row 4 data) by the volume in milliliters (row 8 data). Record your findings in the data table.

5. Calculate the ratio of mass per unit volume for each container by dividing the mass in grams (row 7 data) by the volume in milliliters (row 8 data). Record your results in the data table.

18

6. Make a graph of the mass in grams (row 7 data) and the volume in milliliters (row 8 data) to picture the mass per unit volume ratio found in step 5. Put the volume on the *x*-axis (horizontal axis) and the mass on the *y*-axis (the vertical axis). The mass and volume data from each container will be a data point, so there will be a total of three data points.

7. Draw a straight line on your graph that is as close as possible to the three data points and the origin (0, 0) as a fourth point. If you wonder why (0, 0) is also a data point, ask yourself about the mass of a zero volume of water!

8. Calculate the slope of your graph. (See Appendix II on page 389 for information on calculating a slope.)

9. Calculate your experimental error. Use 1.0 g/mL (grams per milliliter) as the accepted value.

10. Density is defined as mass per unit volume, or mass/volume. The slope of a straight line is also a ratio, defined as the ratio of the change in the *y*-value per the change in the *x*-value. Discuss why the volume data was placed on the *x*-axis and mass on the *y*-axis and not vice versa.

11. Was the purpose of this lab accomplished? Why or why not? (Your answer to this question should show thoughtful analysis and careful, thorough thinking.)

Results

1. What is a ratio? Give several examples of ratios in everyday use.

2. How is the value of π obtained? Why does π not have units?

3. Describe what happens to the surface area to volume ratio for larger and larger cubes. Predict if this pattern would also be observed for other geometric shapes such as a sphere. Explain the reasoning behind your prediction.

4. Why does crushed ice melt faster than the same amount of ice in a single block?

5. Which contains more potato skins: 10 pounds of small potatoes or 10 pounds of large potatoes? Explain the reasoning behind your answer in terms of this laboratory investigation.

6. Using your own words, explain the meaning of the slope of a straight-line graph. What does it tell you about the two graphed quantities?

7. Explain why a slope of mass/volume of a particular substance also identifies the density of that substance.

Problems

An aluminum block that is 1 m × 2 m × 3 m has a mass of 1.62×10^4 kilograms (kg). The following problems concern this aluminum block:

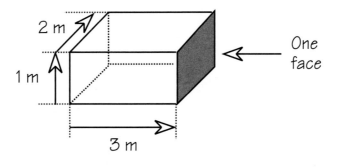

Figure 2.5

1. What is the volume of the block in cubic meters (m^3)?

2. What are the dimensions of the block in centimeters (cm)?

3. Make a sketch of the aluminum block and show the area of each face in square centimeters (cm^2).

4. What is the volume of the block expressed in cubic centimeters (cm^3)?

5. What is the mass of the block expressed in grams (g)?

6. What is the ratio of mass (g) to volume (cm^3) for aluminum?

7. Under what topic would you look in the index of a reference book to check your answer to question 6? Explain.

Data Table 2.1   Circles and Ratios

	Small Circle			Medium Circle			Large Circle		
Trial	1	2	3	1	2	3	1	2	3
Diameter (D)	___	___	___	___	___	___	___	___	___
Circumference (C)	___	___	___	___	___	___	___	___	___
Ratio of C/D	___	___	___	___	___	___	___	___	___

Average $\dfrac{C}{D}$ = _____   Experimental error: _____

Data Table 2.2   Area and Volume Ratios

	Small Cube	Medium Cube	Large Cube
Surface Area	_____	_____	_____
Volume	_____	_____	_____
Ratio of Area/Volume	_____	_____	_____

Data Table 2.3	Mass and Volume Ratios		
Container Number	1	2	3
1. Length of container	_____cm	_____cm	_____cm
2. Width of container	_____cm	_____cm	_____cm
3. Height of container	_____cm	_____cm	_____cm
4. Calculated volume	_____cm^3	_____cm^3	_____cm^3
5. Mass of container	_____g	_____g	_____g
6. Mass of container and water	_____g	_____g	_____g
7. Mass of water	_____g	_____g	_____g
8. Measured volume of water	_____mL	_____mL	_____mL
9. Ratio of calculated volume to measured volume of water	_____cm^3/mL	_____cm^3/mL	_____cm^3/mL
10. Ratio of mass of water to measured volume of water	_____g/mL	_____g/mL	_____g/mL

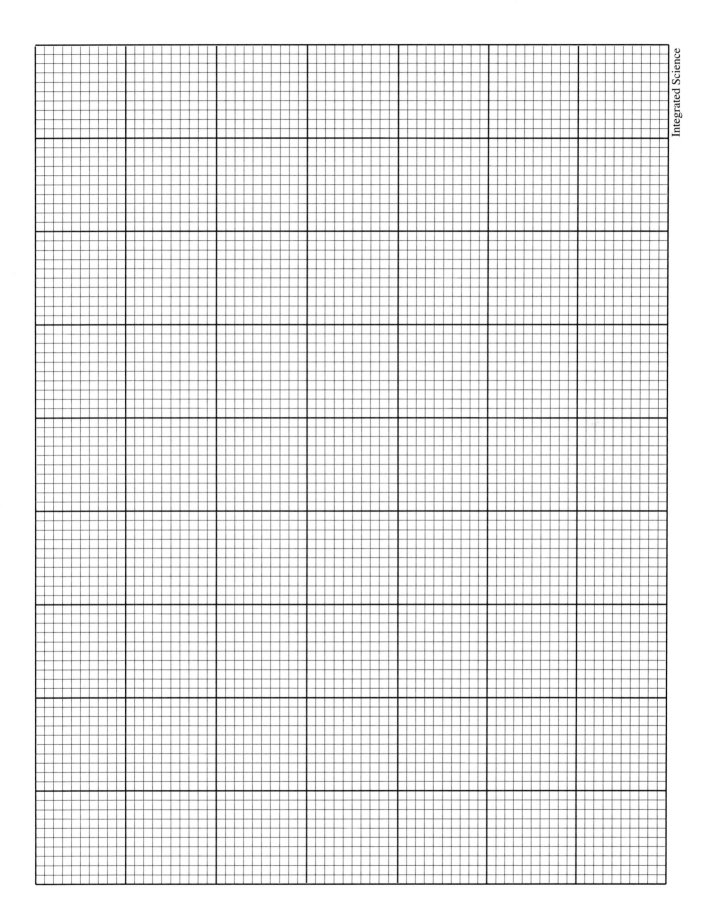

Experiment 3: Motion

Invitation to Inquiry

Have you ever seen an entire stage covered with dominoes lined up, one after another and winding around into interesting patterns? The entertainer tips over one domino, which falls into another, which falls into the one next to it... and on until in a short time all the dominoes have fallen over.

How far apart should the dominoes be spaced for maximum speed? Is it possible to vary this speed by changing the spacing? One domino causes a falling row to continue falling by hitting its neighbor, so the limit to how far apart the dominoes are spaced must be the length of a domino. The other limit would be zero space between two adjacent dominoes, so the limits to the spacing between two adjacent dominoes must be somewhere between zero and one domino length. Thus it would be convenient to record spaces between dominoes as a *ratio* of domino lengths, that is, the space between dominoes in centimeters divided by the length of one domino in centimeters.

You will need to determine how you plan to space the dominoes as well as how many dominoes are needed to measure the speed. By making a graph and doing some calculations, can you predict how many dominoes would be needed—and at what spacing—to make a row that takes exactly 2 minutes to fall?

Ratio of spacing length to domino length (spacing/domino length).

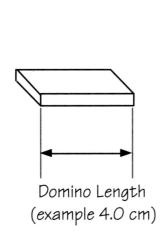

Domino Length
(example 4.0 cm)

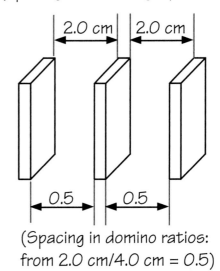

(Spacing in domino ratios:
from 2.0 cm/4.0 cm = 0.5)

Figure 3.1

Background

In this investigation you will analyze and describe motion with a constant velocity and motion with a nonconstant velocity. First, motion with a constant velocity will be investigated by using a battery-operated toy bulldozer or any toy car or truck that moves at a fairly constant speed. Data will be collected and analyzed and a concept will be formalized to described what is happening to the toy as it moves.

Figure 3.2 compares the distance versus time slopes for motion with a constant velocity, with a nonconstant velocity, and with no velocity at all. Note that the slope for some object not moving will be a straight horizontal line. If a vehicle is moving at a uniform (constant) velocity, the line will have a positive slope. This slope will describe the magnitude of the velocity, sometimes referred to as the **speed**. The line for a vehicle moving at a nonconstant speed, on the other hand, will be nonconstant as shown in figure 3.2. A nonconstant speed is also known as accelerated motion, and the ratio of how fast the motion is changing per unit of time is called **acceleration**.

Taking measurable data from a multitude of sensory impressions, finding order in the data, then inventing a concept to describe the order are the activities of science. This investigation applies this process to motion.

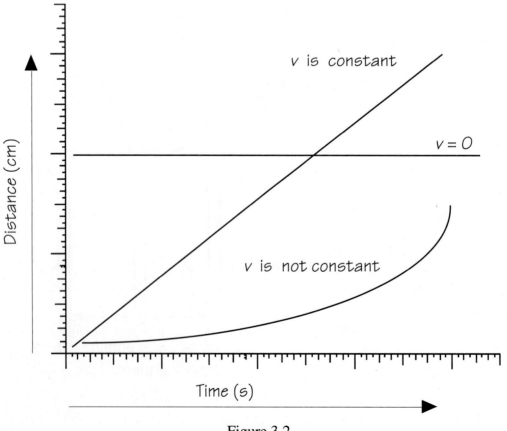

Figure 3.2

28

Procedure

Part A: Constant Velocity on the Level

1. Use masking tape to secure a length of paper such as long sheets of computer paper, rolled butcher paper, or adding machine tape across the floor. The paper should be long enough so the motorized toy vehicle used will not cross the entire length in less than 8 to 10 seconds. Thus, the exact length of paper selected will depend on the vehicle and battery conditions. (Note: Erratic increases or decreases of speed probably mean that a new battery is needed.) The paper will be used to record successive positions of the toy at specific time intervals.

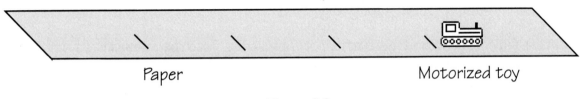

Paper   Motorized toy

Figure 3.3

2. One person with a stopwatch will call out equal time intervals that are manageable but result in at least five or six data points for the total trip. Another person will mark the position of the toy vehicle on the paper when each time interval is called. To avoid interfering with the motion of the toy, mark the position from behind each time. This also means that the starting position should be marked from behind. Other means of measuring velocity that might be used in your laboratory, such as the use of photogates and computer software, will be explained by your instructor.

3. Measure the intervals between the time marks, recording your data in Data Table 3.1 on page 34. Make a graph that describes the motion of the toy vehicle by placing the distance (the dependent variable) on the vertical axis and time (the independent variable) on the horizontal axis. Draw the best straight line as close as possible to the data points. Calculate the slope and record it someplace on the graph.

Part B: Constant Velocity on an Incline

1. This investigation is similar to Part A, but this time the toy vehicle will move up an inclined ramp that is at least 1 m long.

2. Elevate the ramp with blocks or books so that 1 m from the bottom of the ramp is 10 cm high. As in Part A, one person with a stopwatch will call out equal time intervals that are manageable but result in at least five or six data points for the total trip. Another person will mark the position of the toy vehicle on the paper when each time interval is called. To avoid interfering with the motion of the toy, mark the position from behind each time. Also mark the starting position from behind. Measure the intervals between the time marks, recording your data in Data Table 3.2.

29

3. Elevate the ramp to 20 cm high and repeat procedure Part B step 2.

4. Elevate the ramp to 30 cm high and again repeat procedure Part B step 2. Make a graph of all three sets of data in Data Table 3.2. Calculate the slope of each line and write each somewhere on the graph.

Part C: Motion with Nonconstant Velocity

1. You will now set up a track for collecting data about rolling balls. This track can be anything that serves as a smooth, straight guide for a rolling ball. It could be a board with a V-shaped groove, U-shaped aluminum shelf brackets, or two lengths of pipe taped together, for example.

2. The track should be between 1 and 2 m long and supported somewhere between 10 and 50 cm above the table at the elevated end (figure 3.4). In this investigation, a longer track will mean better results. You should consider 1 m as a *minimum* length. Your instructor will describe a different procedure if your lab has photogates, computer software, or different equipment.

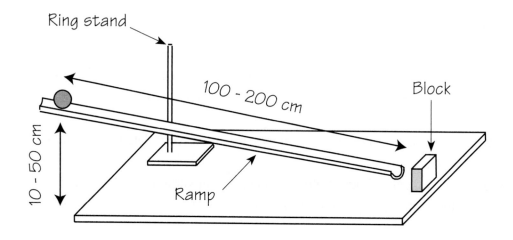

Figure 3.4

3. You will select a minimum of six positions on the ramp from which to release a steel ball or marble. One position should be the uppermost end, and the others should be equally spaced. Hold a ruler across the track with the ball behind it, then release the ball by lifting the ruler straight up the same way each time. Start a stopwatch when the ball is released, then stop it when the ball reaches the bottom of the ramp. A block at the bottom of the ramp will stop the ball and the sound of the ball hitting the block will signal when to stop the stopwatch.

4. Measure the distance and time for three data runs, then average the data for each of the six positions. Record the data in Data Table 3.3 on page 36. Make a graph of the data with time on the *x*-axis.

Results

1. Explain for each part of this investigation how you know if there is or is not a relationship between the variables according to your graphs.

2. For motion with a constant velocity, how do the changes in distance compare for equal time intervals? Is this what you would expect? Explain.

3. What is the rate of travel of the toy over (a) a flat surface, (b) a surface elevated 10 cm high, (c) a surface elevated 20 cm high, and (d) a surface elevated 30 cm high?

4. For motion with a nonconstant velocity, how does the total distance change as the total time increases; that is, do they both increase at the same rate? Explain the meaning of this observation.

5. Considering nonconstant velocity, how do the changes in distance compare for equal time intervals?

6. Was the purpose of this lab accomplished? Why or why not? (Your answer to this question should show thoughtful analysis and careful, thorough thinking.)

Going Further

In part of this investigation, you learned that $\bar{v} = \dfrac{d}{t}$. Using this equation, explain how you can find

(a) the time for a trip when given the average speed and the total distance traveled;

(b) the total distance traveled when given the time for a trip and the average speed; and

(c) the average speed for a trip, no matter what units are used to describe the total distance and the total time of the trip.

Data Table 3.1	Distance and Time Data for Battery-Powered Toy over a Flat Surface
Total Time (s)	Total Distance (cm)
_____	_____
_____	_____
_____	_____
_____	_____
_____	_____
_____	_____

Time (s)	Total Distance (cm)		
	10 cm Elevation	20 cm Elevation	30 cm Elevation
————	————	————	————
————	————	————	————
————	————	————	————
————	————	————	————
————	————	————	————

Data Table 3.2   Distance and Time Data for Battery-Powered Toy over Elevated Surfaces

Data Table 3.3	Time and Distance Data for Rolling Ball on Ramp			
Distance from Bottom (cm)	Time Trial 1 (s)	Time Trial 2 (s)	Time Trial 3 (s)	Time Average (s)
_____	_____	_____	_____	_____
_____	_____	_____	_____	_____
_____	_____	_____	_____	_____
_____	_____	_____	_____	_____
_____	_____	_____	_____	_____
_____	_____	_____	_____	_____

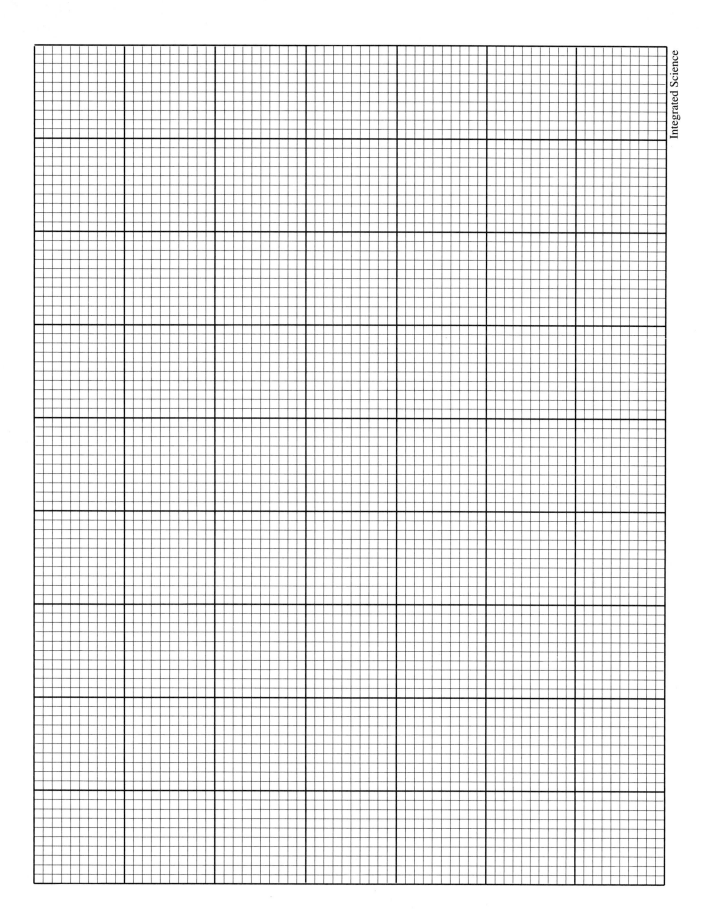

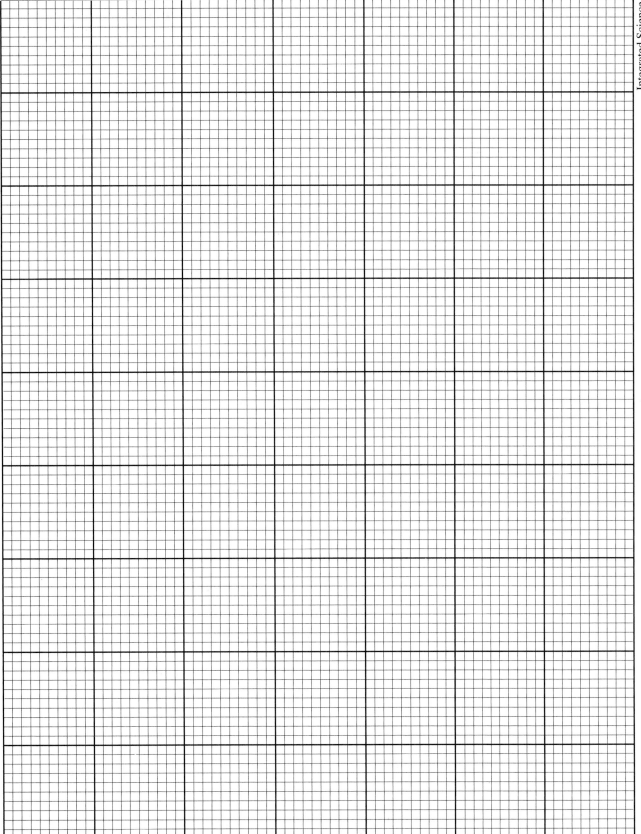

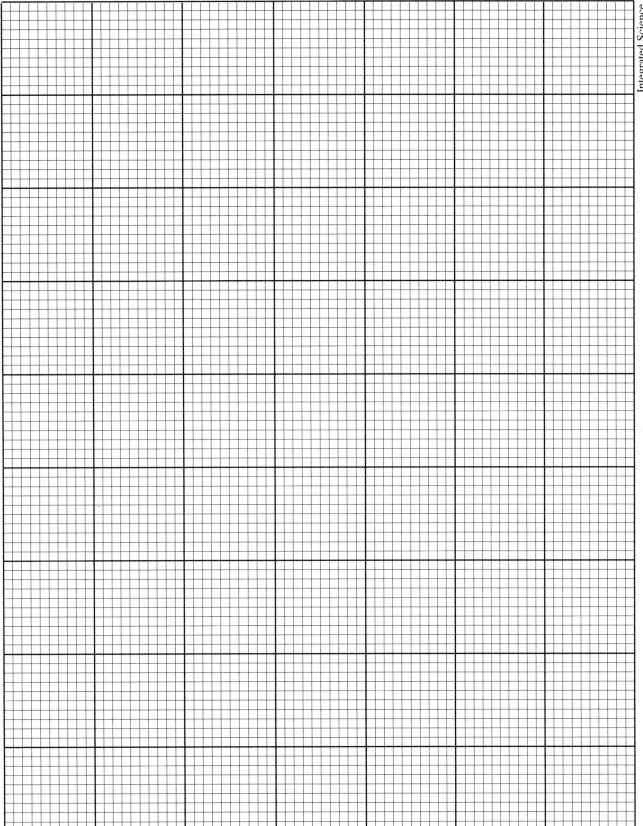

Experiment 4: Free Fall

Invitation to Inquiry

Find out how well you can predict the motion of falling objects. First, select some objects such as a rubber ball, a sheet of notebook paper, and a large metal paper clip. Predict, then study the detail of each object falling independently... for example, what happens to each as they fall? Then compare the motion of the objects side by side. Is it possible to cause them to fall together, at the same time?

Use measurements to construct a graph or graphs that show what is going on between the variables involved in falling objects. Then use the graph to show how to place three or four objects on a long cord. Attach them so when the cord is hung from a high place, then dropped, the objects make a constant plop, plop, plop sound when they hit the ground.

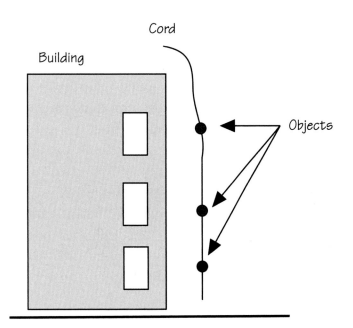

Figure 4.1

Background

In this experiment you will calculate the acceleration of an object as it falls toward the earth's surface. An object in *free fall* moves toward the surface with a uniform accelerated motion due to gravity, g. The value of g varies with location on the surface of the earth, increasing with latitude to a maximum at the poles. The value of g also varies with elevation, decreasing with elevation at a certain latitude. The average, or standard, value of g, however, is usually accepted as 9.8 m/s^2 or 980 cm/s^2.

When you measure the total distance that an object moves during some period of time, you can calculate an average velocity. **Average velocity** is defined as

$$\bar{v} = \frac{\Delta d}{\Delta t}$$

where Δd is the total distance (final distance minus initial, or $d_f - d_i$) and Δt is the total time (final time minus initial, or $t_f - t_i$). In this experiment you will be measuring the velocity of an object that falls from an initial distance and time of zero, so $\Delta d = d_f - 0$ and $\Delta t = t_f - 0$. For the case of a falling object,

$$\bar{v} = \frac{\Delta d}{\Delta t} = \frac{d_f - d_i}{t_f - t_i} \quad \text{since } d_i = 0 \text{ and } t_i = 0 \quad \therefore \quad \bar{v} = \frac{d_f}{t_f}.$$

Thus you can calculate the average velocity of an object in free fall from the total distance traveled and the time of fall. When an object moves with a constant acceleration, you can also find the average velocity by adding the initial and final velocity and dividing by 2,

$$\bar{v} = \frac{v_f + v_i}{2}.$$

By substituting the other expression for average velocity, we have

$$\bar{v} = \frac{v_f + v_i}{2} \quad \text{and} \quad \bar{v} = \frac{d_f}{t_f} \quad \therefore \quad \frac{v_f + v_i}{2} = \frac{d_f}{t_f}.$$

Since the initial velocity of a dropped object is zero, then v_i is zero, and we can solve for the final velocity of v_f, and

$$\frac{v_f}{2} = \frac{d_f}{t_f} \quad \therefore \quad v_f = \frac{2d_f}{t_f}.$$

In this experiment you will measure the distance a mass has fallen during recurring time intervals according to a timing device. This data will enable you to calculate the instantaneous velocity at known time intervals. Plotting the velocity versus the time, then finding the slope will provide an experimental value of g.

Procedure

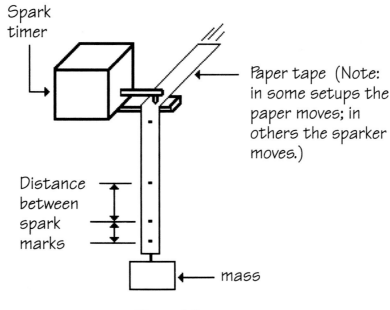

Figure 4.2

1. You will experimentally determine the acceleration due to gravity and compare it to the standard value of 980 cm/s^2. The procedure may vary with the apparatus used. For example, you might use an apparatus that consists of a device to measure the free fall of an object with a spark timer that will mark a paper tape at equal time intervals. As a mass accelerates downward it will leave a trail of spark marks at equal time intervals. You will draw a perpendicular line through each mark, then identify the first mark as your reference line. The first mark is identified as the place where $d_f = 0$. Other means of measuring velocity that might be used in your laboratory, such as the use of photogates and computer software, will be explained by your instructor.

2. For spark mark (or ink dot) trails measure the *total distance* (d_f) by using the beginning mark as a reference line. On page 48, record in Data Table 4.1 the distance in centimeters of each mark *from the reference line*.

3. For each spark, record the *time* (t) that elapsed between the marks as determined by the spark timer. Your instructor will provide exact information for your timer. Most timers are set to operate on 60 Hz, making a spark every 1/60 second. Thus the second spark would have occurred 1/60 second after the first, and the third spark mark would have occurred 1/60 plus 1/60 or 2/60 (0.033 s) after the first mark. Fill in Data Table 4.1 with the total distance and time data for each mark, and calculate and record the velocity at each spark. Repeat the experiment two more times with two more paper tapes, completing Data Table 4.2 (page 49) and Data Table 4.3 (page 50).

Results

1. Look over the data in Data Tables 4.1, 4.2, and 4.3, think about what the data means, then select the data table that seems to have the "best run" data. State which table was chosen and explain the basis for your choice.

2. Using the data table from the best run, make a graph with *velocity* (*v*) on the *y*-axis and *elapsed time* (*t*) on the *x*-axis. (Note: Because the first spark was probably not made at the actual time of release, the line on your graph will probably not have a *y* intercept of 0.) Find the slope and record it here, along with any notes you may wish to record.

3. Use the calculated slope and the accepted value of 980 cm/s^2 to calculate the experimental error.

4. Was the purpose of this lab accomplished? Why or why not? (Your answer to this question should show thoughtful analysis and careful, thorough thinking.)

Going Further

What is your reaction time? One way to measure your reaction time is to have another person hold a meterstick vertically from the top while you position your thumb and index finger at the 50 cm mark. The other person will drop the meterstick (unannounced) and you will catch it with your thumb and finger. Accelerated by gravity (g), the stick will fall a distance (d) during your reaction time (t). Knowing d and g, all you need is a relationship between g, d, and t to find the time.

You know a relationship between d, $\bar{v}$, and t from $\bar{v} = d/t$. Solving for d gives $d = \bar{v}t$.

Any object in free fall, including a meterstick, will have uniformly accelerated motion, so the average velocity is

$$\bar{v} = \frac{v_f + v_i}{2}.$$

Substituting for the average velocity in the previous equation gives

$$d = \left(\frac{v_f + v_i}{2}\right)(t).$$

The initial velocity of a falling object is always zero just as it is dropped, so the initial velocity can be eliminated, giving

$$d = \left(\frac{v_f}{2}\right)(t).$$

Now you want acceleration in place of velocity. From

$$a = \frac{v_f - v_i}{t}$$

and solving for the final velocity gives

$$v_f = at.$$

The initial velocity is again dropped since it equals zero. Substituting the final velocity in the previous equation gives

$$d = \left(\frac{at}{2}\right)(t) \qquad \text{or} \qquad d = \frac{1}{2}at^2.$$

Finally, solving for t gives

$$t = \sqrt{\frac{2d}{g}}.$$

Measuring how far the meterstick falls (in meters) can now be used as the distance (d) with g equaling 9.8 m/s^2 to calculate your reaction time (t).

47

Data Table 4.1	Free Fall Run Number One		
Spark Number	Distance (cm)	Time of Fall (s)	Computed Instantaneous Velocity* (cm/s)
1			
2			
3			
4			
5			
6			
7			
8			
9			
10			

$$*\text{From } v_f = \frac{2d_f}{t_f}$$

Data Table 4.2	Free Fall Run Number Two		
Spark Number	Distance (cm)	Time of Fall (s)	Computed Instantaneous Velocity* (cm/s)
1			
2			
3			
4			
5			
6			
7			
8			
9			
10			

$$*\text{From } v_f = \frac{2d_f}{t_f}$$

Data Table 4.3	Free Fall Run Number Three		
Spark Number	Distance (cm)	Time of Fall (s)	Computed Instantaneous Velocity* (cm/s)
1			
2			
3			
4			
5			
6			
7			
8			
9			
10			

$$*\text{From } v_f = \frac{2d_f}{t_f}$$

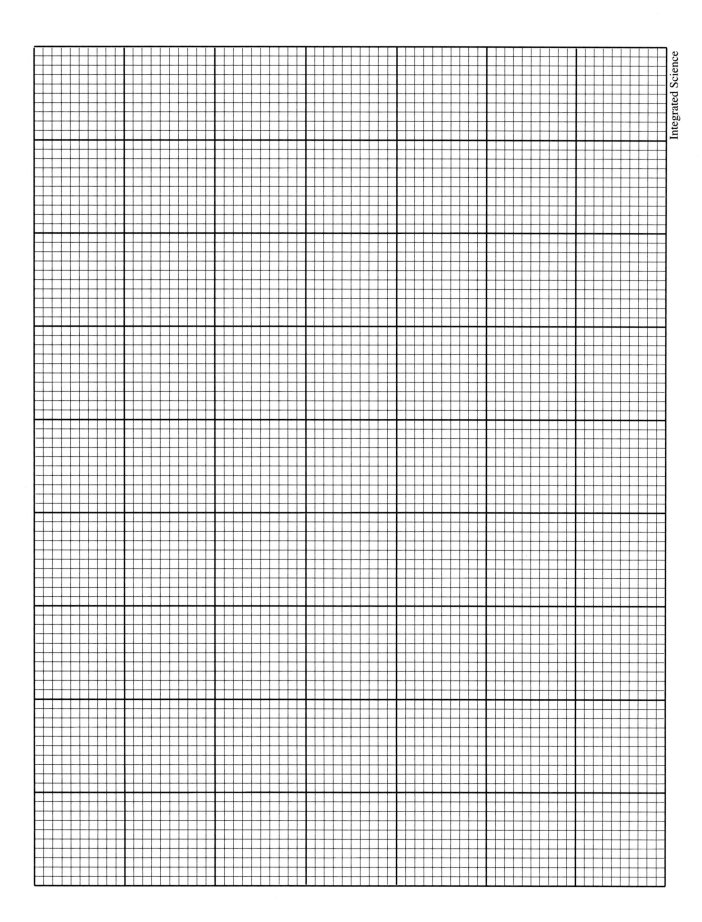

Experiment 5: Centripetal Force

Invitation to Inquiry

1. Did you ever try to figure out which is a cooked egg and which is a raw one without breaking the shell? One way to accomplish this is by spinning the eggs on a plate, and the well-cooked one will continue to spin while the uncooked egg will rock back and forth. The yolk is heavier than the white, but why would an uncooked egg spin more slowly? Use your understanding of centripetal force to develop some ideas about why eggs should behave this way, then design a demonstration or experiment to test your idea.

2. Experiment with some things that rotate, such as rolling cylinders. Roll large, small, solid, hollow, and various combinations of large and solid cylinders and small and solid cylinders down an incline. Predict ahead of time which will reach the bottom of the incline first. Then test your predictions.

3. A hollow and solid cylinder of the same size do not have the same weight. If you roll the two cylinders down an incline slope together, side by side, which cylinder should win? If you attach strings of equal lengths to make pendulums from the same hollow and solid cylinders, will they swing together, side by side? Experiment to find out, then be prepared to explain your findings.

4. Explore relationships between mass and distance from an axis and how hard it is to set an object into rotational motion. Consider using a baton with some kind of movable masses that can be fixed to the baton different distances from the axis of rotation. A large wooden dowel rod and lumps of clay might be a good experimental alternative to a baton.

Background

This experiment is concerned with the force necessary to keep an object moving in a constant circular path. According to Newton's first law of motion there *must be* forces acting on an object moving in a circular path since it does not move off in a straight line. The second law of motion ($F = ma$) also indicates forces since an unbalanced force is required to change the motion of an object. An object moving in a circular path is continuously being accelerated since it is continuously changing direction. This means that there is a continuous unbalanced force acting on the object that pulls it out of a straight-line path. The force that pulls an object out of a straight-line path and into a circular path is called a **centripetal force.**

The magnitude of the centripetal force required to keep an object in a circular path depends on the inertia (or mass) and the acceleration of the object, as you know from the second law ($F = ma$). The acceleration of an object moving in uniform circular motion is $a = v^2/r$, so the magnitude of the centripetal force of an object with a mass (m) that is moving with a velocity (v) in a circular orbit of radius (r) can be found from

$$F = \frac{mv^2}{r}.$$

The distance (circumference) around a circle is $2\pi r$. The velocity of an object moving in a circular path can be found from $v = d/t$ or $v = 2\pi r/T$ where $2\pi r$ is the distance around one complete circle and T is the period (time) required to make one revolution. Substituting for v,

$$F = \frac{m\left(\frac{2\pi r}{T}\right)^2}{r}$$

or

$$F = \frac{\frac{m4\pi^2 r^2}{T^2}}{r},$$

$$F = \frac{4\pi^2 r^2 m}{T^2} \times \frac{1}{r}$$

$$F = \frac{4\pi^2 r\, m}{T^2}.$$

This is the relationship between the centripetal force (F_c), the mass (m) of the object in circular motion, the radius (r) of the circle, and the time (T) required for one complete revolution.

Procedure

1. The equipment setup for this experiment consists of weights (washers) attached to a string and a rubber stopper that swings in a horizontal circle. You will swing the stopper in a circle and adjust the speed so that the stopper does not have a tendency to move in or out, thus balancing the centripetal force (F_c) on the stopper with the balancing force (F_b), or mg, exerted by the washers on the string.

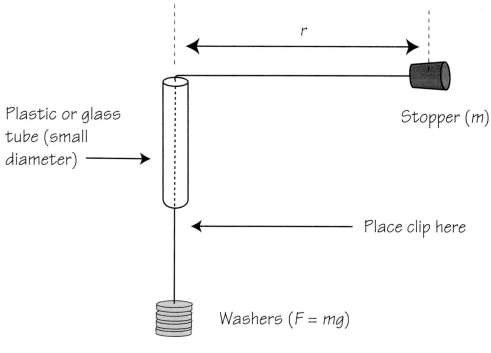

Plastic or glass
tube (small
diameter) ⟶

Stopper (m)

Place clip here

Washers ($F = mg$)

Figure 5.1

2. Place some washers on the string and practice rotating the stopper by placing a finger next to the string, then moving your hand in a circular motion. You are trying to move the stopper with a consistent, balancing motion just enough so the stopper does not move in or out. *Keep the stopper moving in a fairly horizontal circle, without the washers moving up or down.* An alligator (or paper) clip placed on the string just below the tube will help you maintain a consistent motion by providing a point of reference as well as helping with length measurements. Be careful of the moving stopper so it does not hit you in the head.

3. After you have learned to move the stopper with a constant motion in a horizontal plane, you are ready to take measurements. The distance from the string at the top of the tube to the *center* of the stopper is the radius (r) of the circle of rotation. The mass (m) of the stopper is determined with a balance. The balancing force (F_b) of the washers is determined from the mass of the washers times g ($F_b = mg$). The period (T) is determined by measuring the time of a number of revolutions, then dividing the total time by the number of revolutions to obtain the time for one revolution. For example, 20 revolutions in 10 seconds would mean that $^{10}/_{20}$, or 0.5 seconds, is required for one revolution. This data is best obtained by one person acting as a counter speaking aloud while another person acts as a timer.

4. Make four or five trials by rotating the stopper with a different number of washers on the string each time, adding or removing two washers (about 20 g) for each trial. For each trial, record in Data Table 5.1 the mass of the washers, the radius of the circle, and the average time for a single revolution.

Data Table 5.1   Centripetal Force Relationships

Trial	Mass of washers (m)	Balancing force (F_b)	Radius (r)	Time (t)	Centripetal force (F_c)
	(kg)	(N)	(m)	(s)	(N)
1	_____	_____	_____	_____	_____
2	_____	_____	_____	_____	_____
3	_____	_____	_____	_____	_____
4	_____	_____	_____	_____	_____
5	_____	_____	_____	_____	_____

Mass of stopper _____kg

5. Calculate and record the balancing force (F_b) for each trial from the mass of washers times g (9.8 m/s^2), or $F_b = mg$.

6. Calculate and record the centripetal force (F_c) for each trial from

$$F = \frac{4\pi^2 r\, m}{T^2}.$$

Considering the balancing force (F_b) as the accepted value and the calculated centripetal force (F_c) as the experimental value, calculate your percentage error for each trial of this experiment. Analyze the percentage errors and other variables to identify some trends, if any.

Trial 1 :

Trial 2 :

Trial 3 :

Trial 4 :

Trial 5 :

Results

1. Did the balancing force (F_b) equal the centripetal force (F_c)? Do you consider them equal or not equal? Why or why not?

2. Analyze the errors that could be made in all the measured quantities. What was probably the greatest source of error and why? Discuss how these errors could be avoided and how the experiment in general could be improved.

3. Discuss any trends that were noted in your analysis of percentage error for the different trials. Analyze the meaning of any observed trends, or discuss the meaning of the lack of any trends.

4. Was the purpose of this lab accomplished? Why or why not? (Your answer to this question should be reasonable and make sense, showing thoughtful analysis and careful, thorough thinking.)

Experiment 6: Work and Power

Invitation to Inquiry

Tie one end of a string to a book and the other end to a spring scale. Use the spring scale to measure the force needed to pull the book up a smooth board used as a ramp (inclined plane). How much force was required to lift the book straight up (the weight of the book)? How much force was required to pull the book up the ramp? Compare the force needed to lift the book straight up with the force needed to pull the book up the ramp. Is there any relationship of this ratio to the ratio of the length and height of the ramp?

Consider how you can reduce friction on a ramp. Experiment with the use of ball bearings, pencils, or oil between two paper sheets on the surface of the ramp. Compare the results of the force ratios of lifting and pulling the book with the length and height ratio when friction is reduced.

Background

The word *work* represents a concept that has a special meaning in science that is somewhat different from your everyday concept of the term. In science, the concept of mechanical work is concerned with the application of a force to an object and the distance the object moves as a result of the force. Mechanical **work** (W) is defined as the magnitude of the applied force (F) multiplied by the distance (d) through which the force acts, $W = Fd$.

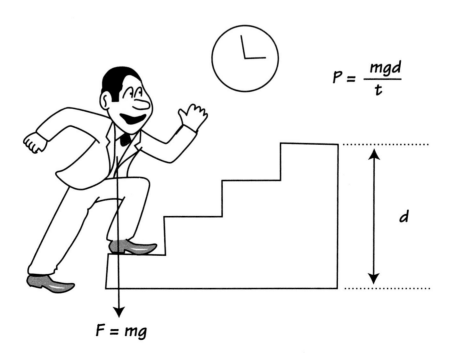

Figure 6.1

You are doing work when you walk up a stairway since you are lifting yourself through a distance. You are lifting your weight (the force exerted) the vertical height of the stairs (distance through which the force is exerted). Running up the stairs rather than walking is more tiring because you use up your energy at a greater rate when running. The rate at which energy is transformed or the rate at which work is done is called power. **Power** (P) is defined as work (W) per unit of time (t),

$$P = \frac{W}{t}$$

When the steam engine was first invented there was a need to describe the rate at which the engine could do work. Since people at that time were familiar with using horses to do their work, the steam engines were compared to horses. James Watt, who designed a workable steam engine, defined **horsepower** (hp) as a power rating of 550 ft·lb/s. In SI units, power is measured in joules per second, called the **watt** (W). It takes 746 W to equal 1 hp, and 1 kW is equal to about 1⅓ hp.

Procedure

1. Teams of two volunteers will measure the work done, the rate at which work is done, and the horsepower rating as they move up a stairwell. Person A will measure and record the data for person B. Person B will measure and record the data for person A. An ordinary bathroom scale can be used to measure each person's weight. Record the weight in pounds (lb) in Data Table 6.1. This weight is the force (F) needed by each person to lift himself or herself up the stairs.

2. The vertical height of the stairs can be found by measuring the height of one step then multiplying by the number of steps in the stairs. Record this distance (d) in feet (ft) in Data Table 6.1.

3. Measure and record the time required for each person to *walk normally* up the flight of stairs. Record the time in seconds (s) in Data Table 6.1.

4. Measure and record the time required for each person to *run* up the flight of stairs as fast as can be safely accomplished. Record the time in seconds (s) in Data Table 6.1.

5. Calculate the work accomplished, power level developed, and horsepower of each person while walking and while running up the flight of steps. Be sure to include the correct units when recording the results in Data Table 6.1.

Results

1. Explain why there is a difference in the horsepower developed in walking and running up the flight of stairs.

2. Is there some limit to the height of the flight of stairs used and the horsepower developed? Explain.

3. Could the horsepower developed by a slower-moving student ever be greater than the horsepower developed by a faster-moving student? Explain.

4. Describe an experiment that you could do to measure the horsepower you could develop for a long period of time rather than for a short burst up a stairwell.

5. Was the purpose of this lab accomplished? Why or why not? (Your answer to this question should show thoughtful analysis and careful, thorough thinking.)

	Volunteer A		Volunteer B	
	Walking	Running	Walking	Running
Weight (*F*) (lb)	_____	_____	_____	_____
Vertical height (*d*) of steps (ft)	_____	_____	_____	_____
Time required (*t*) to *walk* the flight of steps (s)	_____		_____	
Time required (*t*) to *run* the flight of steps (s)		_____		_____
Work done $W = Fd$	_____	_____	_____	_____
Power $P = W/t$	_____	_____	_____	_____
Horsepower developed $P \div 550$ ft·lb/s	_____	_____	_____	_____

Data Table 6.1   Work and Power Data and Calculations

Experiment 7: Thermometer Fixed Points

Invitation to Inquiry

1. Wash an aluminum pop can, leaving a small amount of water in the can. Use tongs to hold the can over a heat source until the water boils, and you can see steam condensing in the air at the opening. Immediately invert the can part way in a container of cool water. Explain what happens in terms of a molecular point of view.

2. Place the ends of a 1m metal rod on two wood blocks and secure one end to its block. Place a pin through a tagboard pointer under the free end. The metal rod should be able to move back and forth, turning the pin as it moves. Explain what happens in terms of a molecular point of view to the pointer as the metal rod is heated then cooled.

3. Boil a small amount of water in a clean 500 mL flask, then apply a round balloon over the mouth before the flask cools. Place a rubber band, doubled if necessary, around the balloon on the neck of the flask. Explain what happens to the balloon as the flask is heated or cooled without using the terms "drawn in" or "suck."

Background

This experiment is concerned with the fixed reference points on the Fahrenheit (T_F) and Celsius (T_C) thermometer scales. Two easily reproducible temperatures are used for the fixed reference points, and the same points are used to define both scales. The fixed points are the temperature of melting ice and the temperature of boiling water under normal atmospheric pressure. The differences in the two scales are (1) the numbers assigned to the fixed points and (2) the number of divisions, called **degrees**, between the two points. On the Fahrenheit scale, the value of 32 is assigned to the lower fixed point, and the value of 212 is assigned to the upper fixed point with 180 divisions between these two points. On the Celsius scale, the value of 0 is assigned to the lower fixed point, and the value of 100 is assigned to the upper fixed point with 100 divisions between these two points. In this laboratory investigation you will compare observed thermometer readings with the actual true fixed points.

Variations in atmospheric pressure have a negligible effect on the melting point of ice but have a significant effect on the boiling point of water. Water boils at a higher temperature when the atmospheric pressure is greater than normal and at a lower temperature when the atmospheric pressure is less than normal. Normal atmospheric pressure, also called **standard barometric pressure**, is defined as the atmospheric pressure that will support a 760 mm column of mercury. An atmospheric pressure change that increases the height of the column of mercury will increase the boiling point by 0.037°C (0.067°F) for each 1.0 mm of additional height. Likewise, an atmospheric pressure change that decreases the height of the column will decrease the boiling point by 0.037°C

(0.067°F) for each 1.0 mm of decreased height. Thus you should add 0.037°C for each 1.0 mm of a laboratory barometer reading above 760 mm and subtract 0.037°C for each 1.0 mm below the normal pressure of 760 mm. This calculation will give you the actual boiling point of water under current atmospheric pressure conditions. Any difference between this value and the observed thermometer reading is an error in the thermometer.

Procedure

1. First, verify accuracy of the lower fixed point of the thermometer. Fill a beaker with cracked ice as shown in figure 7.1. After water begins forming from melting ice, place the bulb end of the thermometer well into the ice, but leave the lower fixed point on the scale uncovered so you can still read it. Gently stir for five minutes and then until you observe no downward movement of the mercury. When you are confident that the mercury has reached its lowest point, carefully read the temperature. The last digit of your reading should be an estimate of the distance between the smallest marked divisions on the scale. Record this observed temperature of the melting point in Data Table 7.1. Use 0°C as the accepted value and calculate and record the measurement error, if any.

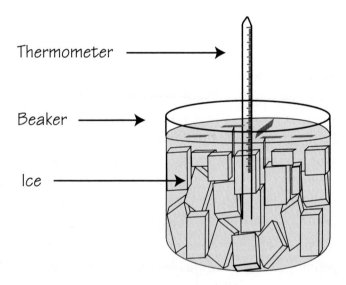

Figure 7.1

2. Now verify the accuracy of the upper fixed point of the thermometer. Set up the steam generator as illustrated in figure 7.2. If you need to insert the thermometer in the stopper, be sure to moisten both with soapy water first. Then hold the stopper with a cloth around your hand and *gently* move the thermometer with a twisting motion. The water in the steam generator should be adjusted so the water level is about 1 cm below the thermometer bulb. When the water begins to boil vigorously, observe the mercury level until you are confident that it has reached its highest point. Again, the last digit of your reading should be an estimate of the distance between the smallest marked divisions on the thermometer scale. Record this observed temperature of the boiling point in Data Table 7.1.

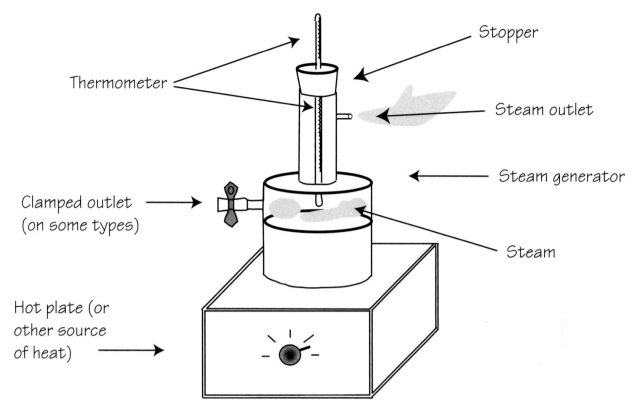

Figure 7.2

3. Determine the accepted value for the boiling point by recording in millimeters the barometric pressure, then calculating the deviation above or below 100°C. Record this accepted boiling point in Data Table 7.1, then calculate and record the measurement error here, if any.

4. Repeat the entire procedure for a second trial, recording all data in Data Table 7.1.

Results

1. Did the temperature change while the ice was melting? Offer an explanation for this observation.

2. Describe how changes in the atmospheric pressure affect the boiling point of water. Offer an explanation for this relationship.

3. Account for any differences observed in the melting point and boiling point readings.

4. How would the differences determined in this investigation influence an experiment concerning temperature if the errors were not considered?

5. Was the purpose of this lab accomplished? Why or why not? (Your answer to this question should show thoughtful analysis and careful, thorough thinking.)

Going Further

Using data from your *best* trial, make a graph by plotting the Celsius temperature scale on the *x*-axis and the Fahrenheit temperature scale on the *y*-axis. Calculate the slope of the straight line and write it here and on the graph somewhere, then answer the following questions.

1. What is the value of the slope? What is the meaning of the slope?

2. What is the value of the *y*-intercept?

3. The slope-intercept form for the equation of a line is $y = mx + b$, where *y* is the variable on the *y*-axis (in this case, °F), *x* is the variable on the *x*-axis (in this case, °C), *m* is the slope of the line, and *b* is the *y*-intercept. Use this information to write the equation of the Celsius-Fahrenheit temperature graph. What is the meaning of this equation?

	Trial 1	Trial 2
Data Table 7.1 Thermometer Readings and Actual Fixed Points		
Observed melting point	_____	_____
Measurement error— melting ice	_____	_____
Observed boiling point	_____	_____
Barometric pressure	_____	_____
Deviation from normal (+ or −)	_____	_____
Accepted boiling point	_____	_____
Measurement error— boiling water	_____	_____

Experiment 8: Specific Heat

Invitation to Inquiry

1. Objects that have been in a room with a constant temperature for some time should all have the same temperature. Touch metal, plastic, and wooden parts of a desk or chair to sense their temperature. Explain your findings.

2. Place about 4 kg of masses on both ends of a small-diameter wire. Place the wire over the center part of a large block of ice that is supported on both ends. Would different kinds of metal wires change the rate of movement of the wire through the ice? Would different thicknesses of wire make a difference?

3. Predict what will happen if you heat brass, glass, and iron balls to 100°C and place them on a sheet of paraffin. Test your prediction, then explain your findings.

Background

Heating is a result of energy transfer, and a quantity of heat can be measured just as any other quantity of energy. The metric unit for measuring energy or heat is the **joule**. However, the separate historical development of the concepts of motion and energy and the concept of heat resulted in separate units. Some of these units are based on temperature differences.

The metric unit of heat is called the **calorie** (cal), a leftover term from the old caloric theory of heat. A calorie is defined as the amount of energy (or heat) needed to increase the temperature of one gram of water one degree Celsius. A kilocalorie (kcal) is the amount of energy (or heat) needed to increase the temperature of one kilogram of water one degree Celsius. The relationship between joules and calories is called the **mechanical equivalence of heat,** and the relationship is

$$4.184 \text{ J} = 1 \text{ cal}$$
$$\text{or}$$
$$4184 \text{ J} = 1 \text{ kcal.}$$

There are three variables that influence the energy transfer that takes place during heating: (1) the temperature change, (2) the mass of the substance being heated, and (3) the nature of the material being heated. The relationships among these variables are as follows:

1. The quantity of heat (Q) needed to increase the temperature of a substance from an initial temperature of T_i to a final temperature of T_f is proportional to $T_f - T_i$, or $Q \propto \Delta T$.

71

2. The quantity of heat (Q) absorbed or given off during a certain ΔT is also proportional to the mass (m) of the substance being heated or cooled, or $Q \propto m$.

3. Differences in the nature of materials result in different quantities of heat (Q) being required to heat equal masses of different substances through the same temperature range.

The **specific heat** (c) is the amount of energy (or heat) needed to increase the temperature of one gram of a substance one degree Celsius. The property of specific heat describes the amount of heat required to heat a certain mass through a certain temperature change, so the units for specific heat are cal/g°C or kcal/kg°C. Note that the k's in the second set of units cancel, so the numerical value for both is the same — for example, the specific heat of aluminum is 0.217 cal/g°C, or 0.217 kcal/kg°C. Some examples of specific heats in these units are:

Aluminum 0.217	Iron 0.113
Copper 0.093	Silver 0.056
Lead 0.031	Nickel 0.106

When the units of all three sets of relationships are the same units used to measure Q, then all the relationships can be combined in equation form,

$$Q = mc\Delta T.$$

This relationship can be used for problems of heating or cooling. A negative result means that energy is leaving a material; that is, the material is cooling. When two materials of different temperatures are involved in heat transfer and are perfectly insulated from their surroundings, the heat lost by one will equal the heat gained by the other,

$$\text{Heat lost}_{\text{(by warm substance)}} = \text{Heat gained}_{\text{(by cool substance)}}$$
$$\text{or}$$
$$Q_{\text{lost}} = Q_{\text{gained}}$$
$$\text{or}$$
$$(mc\Delta T)_{\text{lost}} = (mc\Delta T)_{\text{gained}} \,.$$

Calorimetry consists of using the concept of conservation of energy and applying it to a mixture of materials initially at different temperatures that come to a common temperature. In other words,

$$\text{(heat lost by sample)} = \text{(heat gained by water)}.$$

The sample is heated then placed in water in a calorimeter cup where it loses heat. The water is initially cool, gaining heat when the warmer sample is added. (The role of a Styrofoam calorimeter cup in the heat transfer process can be ignored since two Styrofoam cups have negligible heat gain [$\Delta T \approx 0$] and very little mass.) In symbols,

$$m_s c_s \Delta T_s \;=\; m_w c_w \Delta T_w$$

where m_s is the mass of the sample, c_s the the specific heat of the sample, and ΔT_s is the temperature change for the sample. The same symbols with a subscript w are used for the mass, specific heat, and temperature change of the water. Solving for the specific heat of the sample gives

$$c_s = \frac{m_w c_w \Delta T_w}{m_s \Delta T_s}$$

Procedure

1. You are going to determine the specific heat of three samples of different metals by using calorimetry. You will run two trials on each sample, making *very careful* temperature and mass measurements. Do the calculations before you leave the lab. If you have made a mistake you will still have time to repeat the measurements if you know this before you leave.

2. Be sure you have sufficient water to cover at least the bottom two-thirds of a submerged metal boiler cup (see figure 8.1) but not so much water that it could slosh into the cup when the water is boiling. Start heating the water to a full boil as you proceed to the next steps.

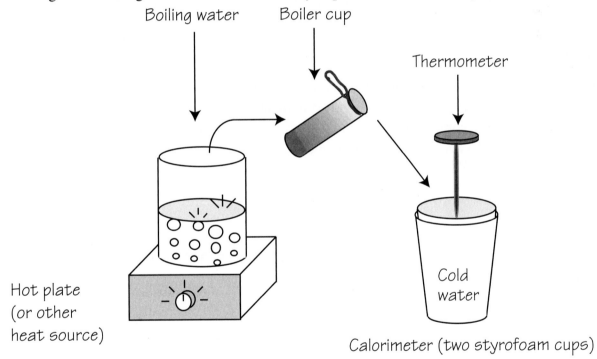

Figure 8.1

3. Measure and record the mass of a dry boiler cup. Pour metal shot into the boiler cup until it is about one-third filled, then measure and record the mass of the cup plus shot. Record the mass of the metal sample (m_s) in Data Table 8.1.

4. Carefully insert a thermometer into the metal shot, positioning it so the sensing end is in the middle of the shot not touching the sides of the boiler cup. Carefully lower the boiler cup into the boiling water. Heat the metal shot until it is in the range of 90° to 95°C. Allow the sample to continue heating as you prepare the water and calorimeter cup (steps 5 and 6).

5. Acquire or make a calorimeter cup of two Styrofoam cups, one placed inside the other (figure 8.1) to increase the insulating ability of the cup. Measure and record the mass of the two cups. Add just enough water to the cup to cover the metal shot when it is added to the cup. This water should be cooler than room temperature (this is to balance possible heat loss by radiation). Measure and record the initial temperature of the water (T_{iw}) in Data Table 8.1.

6. Determine the mass of the cup with the water in it, then subtract the mass of the cup to find the mass of the cold water (m_w). Record the mass of the cold water in Data Table 8.1.

7. Measure and record the temperature of the metal shot. Record the initial temperature of the sample (T_{is}) in Data Table 8.1.

8. Pour the metal shot into the the water in the Styrofoam calorimeter cup. Stir and measure the temperature of the mixture until the temperature stabilizes. Record this stabilized temperature and the final temperature for the water (T_{fw}) and the final temperature for the metal sample (T_{fs}). Calculate the specific heat (c_s) of the metal sample. Note that ΔT_w is obtained from $|T_{fw} - T_{iw}|$ and ΔT_s is obtained from $|T_{fs} - T_{is}|$.

8. Repeat the above steps for sample 2, recording all measurement data in Data Table 8.2. Repeat the procedure for sample 3, recording all measurement data in Data Table 8.3. Run a second trial on all three samples, comparing the results of both trials on each sample. Compare the calculations from the two trials on each sample to decide if a third trial is needed.

Results

1. Calculate the specific heat (c_s) for each sample. Show all work here and record your result in each data table.

2. Using the accepted value for each sample, calculate the percentage error here and record it in each data table.

3. Discuss and evaluate the magnitude of various sources of error in this experiment.

4. What would happen to the calculated specific heat if some boiling water were to slosh into the cup with the metal?

5. Was the purpose of this lab accomplished? Why or why not? (Your answer to this question should show thoughtful analysis and careful, thorough thinking.)

Data Table 8.1   Specific Heat of _____

	Trial 1	Trial 2
Mass of sample (m_s)		
Initial temperature of cold water (T_{iw})		
Mass of cold water (m_w)		
Initial temperature of metal sample (T_{is})		
Final temperature of metal sample (T_{fs})		
Final temperature of water (T_{fw})		
Calculated specific heat (c_s)		
Accepted value		
Percent error		

Data Table 8.2   Specific Heat of _____

	Trial 1	Trial 2
Mass of sample (m_s)	_____	_____
Initial temperature of cold water (T_{iw})	_____	_____
Mass of cold water (m_w)	_____	_____
Initial temperature of metal sample (T_{is})	_____	_____
Final temperature of metal sample (T_{fs})	_____	_____
Final temperature of water (T_{fw})	_____	_____
Calculated specific heat (c_s)	_____	_____
Accepted value	_____	_____
Percent error	_____	_____

Data Table 8.3   Specific Heat of _____

	Trial 1	Trial 2
Mass of sample (m_s)	_____	_____
Initial temperature of cold water (T_{iw})	_____	_____
Mass of cold water (m_w)	_____	_____
Initial temperature of metal sample (T_{is})	_____	_____
Final temperature of metal sample (T_{fs})	_____	_____
Final temperature of water (T_{fw})	_____	_____
Calculated specific heat (c_s)	_____	_____
Accepted value	_____	_____
Percent error	_____	_____

Experiment 9: Speed of Sound in Air

Invitation to Inquiry

1. For any sound there is a relationship between v, f, and λ. For any sound produced in a closed air column there is also a relationship between the temperature, λ, and the length of the shortest air column at which resonance occurs. Therefore, it should be possible to calibrate a closed air column, making marks on the side of the tube so you can use it as a thermometer. How can you make a sound-resonance thermometer that will show the present temperature?

2. Wash your hands thoroughly with soap and water and dry them, paying particular attention to cleaning and drying the forefinger of your writing hand. Dip your clean forefinger in a half-filled, thin-walled water glass. Slowly run your forefinger around the top of the rim of the glass. You might need to dry your finger then start over several times, but eventually if you keep after it a faint, somewhat shrill continuous ringing note will be produced. Your clean, wet finger makes many minute catches on the glass rim as you move it round and round. The ringing note comes from the many tiny catches of your finger. All the tiny forces from your moving finger cause the glass to vibrate, and the vibrating glass produces the continuous note. The pitch of the note will depend on the glass used and the amount of water in the glass. The vibration can be seen clearly on the water surface as they establish circular standing waves.

3. The ringing, rubbed water glass described above can be used to study resonance. First, obtain two similar thin-walled glasses. The two glasses should make the same note when tapped. If they do not make the same pitched notes, add water to one until they do. Place the two water-adjusted glasses side by side about 3 cm apart. Rub your freshly washed finger slowly round the rim of one of the glasses, being careful not to disturb the other. Observe the second (unrubbed) glass as the humming note is produced from the first glass. The second glass should start to vibrate with the first. Experiment with water levels, different distances between the two glasses, and other variables that might influence the resonance condition. Report your finding of the optimum condition for resonance.

Background

A vibrating tuning fork sends a series of condensations and rarefactions through the air. When the tuning fork is held over a glass tube that is closed at the bottom, the condensations and rarefactions are reflected from the bottom. At certain lengths of tube, the reflected condensations and rarefactions are in phase with those being sent out by the tuning fork, and an increase of amplitude occurs from the resonant condition. Figure 9.1 shows a wave trace representing one wavelength in which the reflected wave is in phase with the incoming wave forming a standing wave. The antinodes represent places of maximum vibration and increased amplitude.

Incoming wave + Reflected wave = Standing wave

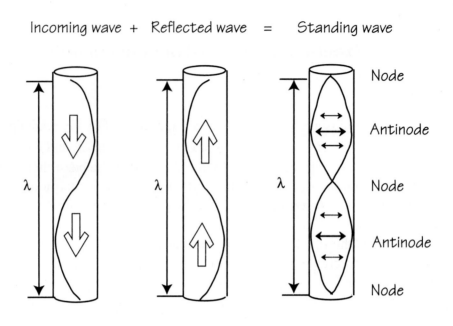

Figure 9.1

Resonance occurs when the length of the tube is such that an antinode (the place of maximum vibration) occurs at the open end. As you can see from the sketch above, there are two situations when this would occur for tube lengths less than one wavelength, 1/4 of the way up and 3/4 of the way up from the bottom. Thus resonance occurs when the length of the tube (L) is equal to 1/4 λ, 3/4 λ, 5/4 λ, and so forth where λ is the wavelength of the sound wave produced by the tuning fork.

In this experiment, a vibrating tuning fork is held just above a cylinder that is open at one end. The length to the closed end is adjusted by adding or removing water. The lowest frequency (the fundamental frequency) occurs when the longest wavelength has an antinode at the open end, so the length of the open tube is about 1/4 of the wavelength of the fundamental frequency as shown in figure 9.2. Since the length of the tube at this fundamental frequency is $L = 1/4\ \lambda$, then the fundamental wavelength must be $\lambda = 4L$.

Using the wave equation

$$v_T = f\lambda,$$

and substituting the known frequency of the tuning fork for f and the experimentally determined value for the wavelength λ, you can calculate the speed of sound v_T in the tube at room temperature by using the relationship

$$v_T = v_{0°C} + \left(\frac{0.6\ \text{m}/\text{s}}{°C}\right)(T_{\text{room}}),$$

where $v_{0°C}$ is the speed of sound at 0° C (331.4 m/s), and T_{room} is the present room temperature in °C.

Procedure

1. The water level in the glass tube is adjusted by raising and lowering the supply tank. Adjust the tank so the glass tube is nearly full of water.

2. Strike the tuning fork with a rubber hammer and hold the vibrating tines just above the opening of the tube.

3. Lower the water level slowly while listening for the increase in the intensity of the sound that comes with resonance. Experiment with the *entire length of the tube*, seeing how many different places of resonance you can identify.

4. Using the information learned in procedure step 3, go to the resonance level immediately *below the resonance position* of the highest water level as shown in figure 9.2. (Make sure there is *not* another resonance point between the highest water level and this second level.) Slightly raise and lower the water level until you are sure that you have found the maximum intensity. Note the relationship between the wavelength and the length of the tube as shown in figure 9.2. Measure and record in Data Table 9.1 on page 84 the length of this resonating air column to the nearest millimeter. Change the water level and run two more trials, again locating the distance with the maximum sound. Record these two lengths in Data Table 9.1 and average the length for the three trials. Record the frequency of the tuning fork (usually stamped on the handle) and the room temperature.

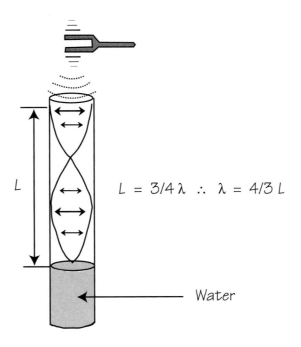

$$L = 3/4\,\lambda \quad \therefore \quad \lambda = 4/3\,L$$

Water

Figure 9.2

5. Repeat procedure steps 1 through 4 for the second resonance point *at the highest water level* with an air column about one-third the length of the first as shown in figure 9.3. (Again, make sure there is *not* another resonance point between the highest water level and this second level.) Note the relationship between the wavelength and the length of the tube as shown in figure 9.3. Run three trials at this position and record the data in Data Table 9.2 on page 84 and, as before, average the three trials. Record the frequency of the tuning fork and the room temperature (do not assume that the room temperature remains constant).

6. Repeat the entire procedure using a different tuning fork with a different frequency. Record all data in Data Tables 9.3 and 9.4 on page 85.

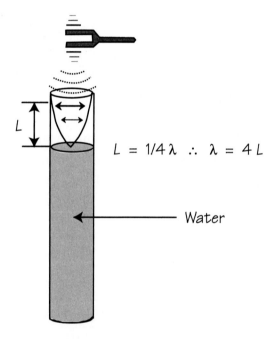

$$L = 1/4 \lambda \quad \therefore \quad \lambda = 4L$$

Water

Figure 9.3

Results

1. Calculate the velocity of sound at room temperature for both tuning forks at both resonance positions and record in data tables at the measured room temperatures. Write the average values for both tuning forks here.

2. Using the accepted value of sound in dry air at the measured room temperature, calculate the percentage error for both tuning forks [accepted value = 331.4 m/s + (0.6 m/s/°C)(T_{room})].

3. Analyze and discuss the possible sources of error in this experiment.

4. Describe how you could do a similar experiment to find the frequency of a tuning fork with an unknown frequency.

5. Was the purpose of this lab accomplished? Why or why not? (Your answer to this question should show thoughtful analysis and careful, thorough thinking.)

Data Table 9.1	Resonance in an Air Column: Lowest Position - Frequency 1			
	Trial 1	Trial 2	Trial 3	Average
Length of resonating air column (m)				
Room temperature (°C)				
Calculated Wavelength (m)				
Calculated velocity in air (m/s)				
Tuning fork frequency (Hz)...				

Data Table 9.2	Resonance in an Air Column: Next Higher Position - Frequency 1			
	Trial 1	Trial 2	Trial 3	Average
Length of resonating air column (m)				
Room temperature (°C)				
Calculated Wavelength (m)				
Calculated velocity in air (m/s)				
Tuning fork frequency (Hz)...				

84

Data Table 9.3	Resonance in an Air Column: Lowest Position - Frequency 2			
	Trial 1	Trial 2	Trial 3	Average
Length of resonating air column (m)				
Room temperature (°C)				
Calculated Wavelength (m)				
Calculated velocity in air (m/s)				
Tuning fork frequency (Hz)..				

Data Table 9.4	Resonance in an Air Column: Next Higher Position - Frequency 2			
	Trial 1	Trial 2	Trial 3	Average
Length of resonating air column (m)				
Room temperature (°C)				
Calculated Wavelength (m)				
Calculated velocity in air (m/s)				
Tuning fork frequency (Hz)..				

Experiment 10: Static Electricity

Invitation to Inquiry

1. This inquiry experiment works best on a day with low humidity. Tie a string around the lip of a small, glass test tube. Try rubbing two tubes with different kinds of cloth as you then allow the tubes to hang freely near each other and observe any interactions.

2. Try rubbing a hard plastic comb with fur or flannel for several minutes. Bring the comb near a hanging test tube that has been rubbed with cloth and observe any interactions.

3. Try rubbing two combs with fur or flannel for several minutes. Bring one comb near a hanging comb that has been rubbed with fur or flannel and observe any interactions.

4. Extend your investigation to other materials or objects if you wish. Explain the meaning of what you find in your experiments with the test tube, combs, and other materials.

Background

Charges of static electricity are produced when two dissimilar materials are rubbed together. Often the charges are small or leak away rapidly, especially in humid air, but they can lead to annoying electrical shocks when the air is dry. The charge is produced because electrons are moved by friction, and this can result in a material acquiring an excess of electrons and becoming a negatively charged body. The material losing electrons now has a deficiency of electrons and is a positively charged body. All electric static charges result from such gains or losses of electrons. Once charged by friction, objects soon return to the neutral state by the movement of electrons. This happens more quickly in humid air because water vapor assists with the movement of electrons from charged objects. In this experiment you will study the behavior of static electricity, hopefully on a day of low humidity.

Procedure

Part A: Attraction and Repulsion

1. Rub a glass rod briskly for several minutes with a piece of nylon or silk. Suspend the rod from a thread tied to a wooden meterstick as shown in figure 10.1. Rub a second glass rod briskly for several minutes with nylon or silk. Bring it near the suspended rod and record your observations in Data Table 10.1. (If nothing is observed to happen, repeat the procedure and rub both rods briskly for twice the time.)

2. Repeat the procedure with a hard rubber rod that has been briskly rubbed with wool or fur. Bring a second hard rubber rod that has also been rubbed with wool or fur near the suspended rubber rod. Record your observations as in procedure step 1.

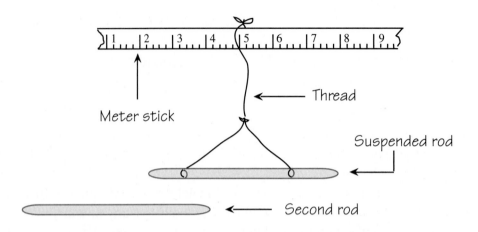

Figure 10.1

3. Again rub the hard rubber rod briskly with wool or fur and suspend it. This time briskly rub a glass rod with nylon or silk and bring the glass rod near the suspended rubber rod. Record your observations.

4. Briskly rub a glass rod with nylon or silk and bring it near, but not touching, the terminal of an electroscope (figure 10.2). Record your observations.

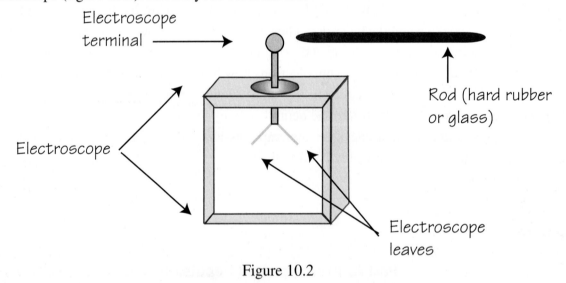

Figure 10.2

5. Repeat procedure step 4 with a hard rubber rod rubbed with wool or fur, again not touching the electroscope terminal. Record your observations.

Part B: Charging by Induction

1. Inflate two rubber balloons and tie the ends. Attach threads to each balloon and hang them next to each other from a support. Rub both balloons with fur or wool and allow them to hang freely. Record your observations in Data Table 10.2.

88

2. Bring a hard rubber rod that has been rubbed with wool or fur near the rubbed balloons. Record your observations.

3. Bring a glass rod that has been rubbed with nylon or silk near the rubbed balloons. Record your observations.

4. Detach one of the balloons by breaking or cutting the thread. Rub the balloon with fur or wool for several minutes. Hold the balloon against a wall and slowly release it. Record your observations.

5. Move the rubbed balloon near an electroscope and record your observations.

6. Move an electroscope near the wall where the balloon was held. Record your observations.

Part C: Determining the Sign of a Charge

1. When a rubbed hard rubber rod is brought near the terminal of an electroscope the leaves will stand apart but fall back together when the rod is removed.

2. When a rubbed hard rubber rod touches the terminal of an electroscope the leaves stand apart as before. When the rod is removed this time the leaves *remain* apart.

3. When the charged rod was brought near the terminal a charge was *induced* by the reorientation of charges in the terminal and leaves. When the rod was removed, the charges returned to their original orientation and the leaves collapsed because no net charge remained on the electroscope.

4. When the electrode was touched, charge was transferred to (or from) the electroscope and removing the rod had no effect on removing the charge. Touching the terminal with your finger returns the electroscope to a neutral condition.

5. An electroscope may be used to determine the sign of a charged object. First, charge the electroscope by induction as in procedure step 3 above. While the charged rod is near the terminal, touch the opposite side of the terminal with a finger of your free hand. Electrons will be repelled and conducted away through your finger. Remove your finger from the terminal, then move the rubber rod from near the electroscope. The electroscope leaves now have a net positive charge. If a charged object is brought near the electroscope, the leaves will spread farther apart if the object has a positive charge. If the charged object has a negative charge, electrons are repelled into the leaves and they will move together as they are neutralized.

6. The process of an object gaining an excess of electrons or losing electrons through friction is complicated and not fully understood theoretically. It is possible experimentally, however, to make a list of materials according to their ability to lose or gain electrons. Gather various materials such as polyethylene film, rubber, wood, cotton, silk, nylon, fur, wool, glass, and plastic. Give an

electroscope a positive charge by induction as described in procedure Part C step 5. Rub combinations of the materials together and determine if the charge on each material is positive or negative. Record your findings.

Results

1. Describe two different ways that electrical charge can be produced by friction.

2. Describe how you can determine the sign of a charged object. What assumption must be made using this procedure?

3. Move a hard rubber rod that has been rubbed with wool or fur near a very thin, steady stream of water from a faucet. Describe, then explain your observations.

4. Was the purpose of this lab accomplished? Why or why not? (Your answer to this question should be reasonable and make sense, showing thoughtful analysis and careful, thorough thinking.)

Data Table 10.1	Attraction and Repulsion of Glass Rod and Rubber Rod
Interaction	Observations
Glass rod - Glass rod	
Rubber rod - Rubber rod	
Glass rod - Rubber rod	
Glass rod - Electroscope	
Rubber rod - Electroscope	

How many kinds of electric charge exist according to your findings above? Explain your reasoning.

How do charges interact?

Data Table 10.2	Charging by Induction
Interaction	Observations
Balloon - Balloon	
Rubber rod - Balloon	
Glass rod - Balloon	
Balloon - Wall	
Balloon - Electroscope	
Wall - Electroscope	

What evidence did you find to indicate that the balloons had static charges?

Describe the evidence you found to indicate that the wall was or was not charged as shown by the electroscope. Explain.

Explain why a balloon exhibits the behavior that it did on the wall.

Experiment 11: Ohm's Law

Invitation to Inquiry

The carbon resistors that are used as standard sources of resistance in electrical circuits are marked with a code of colored bands. Here is the code for the colors:

Black = 0	Green = 5
Brown = 1	Blue = 6
Red = 2	Violet = 7
Orange = 3	Gray = 8
Yellow = 4	White = 9

The value of the resistor is $AB \times 10^C \pm D$ where no D band means $\pm 20\%$, silver means $\pm 10\%$, and

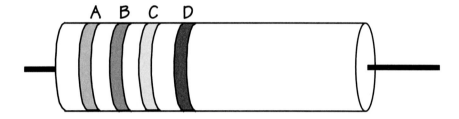

gold means $\pm 5\%$. The band placement is shown above. As an example, consider bands of red, yellow, red, and silver on a resistor. This means $24 \times 10^2 \pm 10\%$ ohms, or $2400 \pm 240 \ \Omega$.

Obtain 5 or 6 resistors and a meter to measure the experimental resistance of each. Read the code to determine the accepted value, then find the experimental error as described in the Appendix III. What could account for experimental errors, if any?

Background

An electric charge has an electric field surrounding it, and work must be done to move a like-charged particle into this field since like charges repel. The electrical potential energy is changed just as gravitational potential energy is changed by moving a mass in the earth's gravitational field. A charged particle moved into the field of a like-charged particle has potential energy in the same way that a compressed spring has potential energy. In electrical matters the potential difference that is created by doing work to move a certain charge creates electrical potential. A measure of the electrical potential difference between two points is the **volt** (V).

A volt measure describes the potential difference between two places in an electric circuit. By analogy to pressure on water in a circuit of water pipes the potential difference is sometimes called an "electrical force" (emf). Also by analogy to water in a circuit of water pipes, there is a varying rate of flow at various pressures. An electric **current** *(I)* is the quantity of charge moving through a conductor in a unit of time. The unit defined for measuring this rate is the **ampere** *(A)*, or the **amp** for short.

The rate of water flow in a pipe is directly proportional to the water pressure; for example, a greater pressure produces a greater flow. In an electric circuit the current is directly proportional to the potential difference (V) between two points. Most materials, however, have a property of opposing or reducing a current, and this property is called **electrical resistance** *(R)*. If a conductor offers a small resistance, less voltage would be required to push an amp of current through the circuit. On the other hand, a greater resistance requires more voltage to push the same amp of current through the circuit. Resistance *(R)* is therefore a *ratio* of the potential difference (V) between two points and the resulting current. This ratio is the unit of resistance and is called an **ohm** (Ω). Another way to show the relationship between the voltage, current, and resistance is

$$R = \frac{V}{I}$$

or

$$V = IR$$

which is known as **Ohm's law**. This is one of the three ways to show the relationship; this one (solved for V) happens to be the equation of a straight line with a slope *R* when V is on the *y*-axis, *I* is on the *x*-axis, and the *y*-intercept is zero.

Procedure

Part A: Known Resistance

1. A known resistance will be provided for use in this circuit:

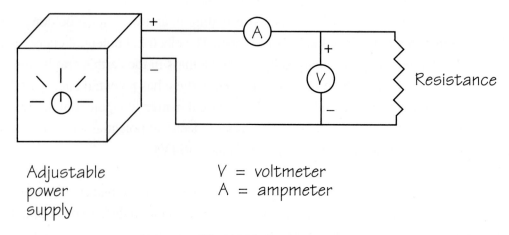

Figure 11.1

2. You will adjust the dc adjustable power supply as instructed by your laboratory instructor, obtaining six values for voltage and current using the supplied resistor. Set up the circuit with the power *off* and do not proceed until the laboratory instructor has checked the circuit.

3. Record the value of the resistor and the six values for the current and voltage in Data Table 11.1.

Part B: Unknown Resistance

Repeat procedure A with an unknown resistor. Record your data in Data Table 11.2.

Results

1. Make a graph of the six data points of Data Table 11.1, placing the current on the *x*-axis and the voltage on the *y*-axis. Calculate the slope and write it here and somewhere on the graph.

2. Compare the calculated value of the known resistor with the accepted value as given by your instructor. Calculate the percentage error.

3. Make a second graph, this time of the six data points in Data Table 11.2, again placing the current on the *x*-axis and the voltage on the *y*-axis. Calculate the slope and write it here and somewhere on the graph.

4. What is the value of the unknown resistor?

5. Explain how the two graphs demonstrate Ohm's law.

6. Was the purpose of this lab accomplished? Why or why not? (Your answer to this question should show thoughtful analysis and careful, thorough thinking.)

Going Further

1. Check your answer about the value of the unknown resistor by using your calculated value in the equation of a straight line when V = 2 V, 4 V, and 6 V. Verify with the laboratory equipment and calculate the average percentage error. Describe your results here.

2. Use three different resistances (e.g., 16Ω, 30 Ω, and 47 Ω) connected in a series for four different input voltages (2 V, 4 V, 6 V, and 8 V) and connected in a parallel circuit. Plot voltage versus total current for both the series and parallel circuits and quantitatively show how the total resistance (the slope) differs for series and parallel circuits.

Data Table 11.1	Voltage and Current Relationships With Known Resistance	
Trial	Voltage (*V*)	Current (*I*)
1	_____	_____
2	_____	_____
3	_____	_____
4	_____	_____
5	_____	_____
6	_____	_____

Resistor _____Ω

Data Table 11.2	Voltage and Current Relationships With Unknown Resistance	
Trial	Voltage (*V*)	Current (*I*)
1	_____	_____
2	_____	_____
3	_____	_____
4	_____	_____
5	_____	_____
6	_____	_____

Resistor _____ Ω

Integrated Science

Experiment 12: Magnetic Fields

Invitation to Inquiry

Is it possible to produce electricity in an extension cord that is not plugged into a circuit? Hook a 50 ft extension cord to a galvanometer and move it as a jump rope, cutting the magnetic field lines around the earth. Figure out how you are going to attach the cord to the galvanometer and how you are going to move it across the earth's magnetic field lines. Can you think of any practical uses for "jump-rope electricity?"

Background

A magnet moved into the space near a second magnet experiences a force as it enters the **magnetic field** of the second magnet. The magnetic field model is a conceptual way of considering how two magnets interact with one another. The magnetic field model does not consider the force that one magnet exerts on another one through a distance. Instead, it considers *the condition of space* around a magnet. The condition of space around a magnet is considered to be changed by the presence of the magnet. Since this region of space, or field, is produced by a magnet, it is called a *magnetic field*. A magnetic field can be represented by *magnetic field lines*. By convention, magnetic field lines are drawn to indicate how the *north pole* of a tiny imaginary magnet would point when in various places in the magnetic field. Arrowheads indicate the direction that the north pole would point, thus defining the direction of the magnetic field. The strength of the magnetic field is greater where the lines are closer together and weaker where they are further apart. Magnetic field lines emerge from a magnet at the north pole and enter the magnet at the south pole. Magnetic field lines always form closed loops.

Magnetic field strength is defined in terms of the magnetic force exerted on a test charge of a particular charge and velocity. The magnetic field is thus represented by vectors (symbol B) which define the field strength, also called the magnetic induction. The units are:

$$B = \frac{\text{newton}}{(\text{coulomb})\left(\dfrac{\text{meters}}{\text{second}}\right)}.$$

Since a coulomb/second is an amp, this can be written as

$$B = \frac{\text{newton}}{\text{amp} \cdot \text{meter}}$$

which is called a **tesla** (T). The tesla is a measure of the strength of a magnetic field. Near the surface, the earth's horizontal magnetic field in some locations is about 2×10^{-5} tesla. A small bar magnet produces a magnetic field of about 10^{-2} tesla, but large, strong magnets can produce magnetic fields of 2 tesla. Superconducting magnets have magnetic fields as high as 30 tesla. Another measure of magnetic field strength is called the **gauss** (*G*) (1 tesla = 10^4 gauss). Thus the process of demagnetizing something is sometimes referred to as "degaussing."

In this experiment you will investigate the magnetic field around a permanent magnet.

Procedure

1. Tape a large sheet of paper on a table with the long edge parallel to the north-south magnetic direction as determined by a compass.

2. Center a bar magnet on the paper with its south pole pointing north. Use a sharp pencil to outline lightly the bar magnet on the paper. Write N and S on the paper to record the north-seeking and south-seeking poles of the bar magnet. Place the bar magnet back on its outline if you moved it to write the N and the S.

3. Slide a small magnetic compass across the paper, stopping close to the north-seeking pole of the bar magnet. Make two dots on the paper, one on either side of the compass and aligned with the compass needle. See figure 12.1.

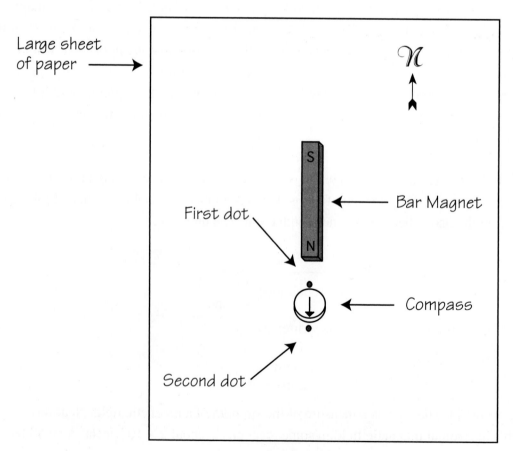

Figure 12.1

4. Slide the compass so the south pole of the needle is now directly over the dot that was at the north pole of the needle. Make a new dot at the north pole end of the compass, exactly in front of the needle. See figure 12.2.

5. Continuing the process of moving the compass so the south pole of the needle is over the most recently drawn dot, then make another new dot at the north pole of the needle. Stop when you reach the bar magnet or the edge of the paper.

6. Draw a smooth curve through the dots, using several arrowheads to show the direction of the magnetic flux line.

7. Repeat procedure steps 3 through 6 by starting with the compass in a new location somewhere around the bar magnet. Repeat the procedures until enough flux lines are drawn to make a map of the magnetic field.

8. Place a second large sheet of paper on a large rigid plastic sheet (or glass plate) on top of the bar magnet. Sprinkle a thin, even layer of iron filings over the plastic, tapping the sheet lightly as you sprinkle. Sketch the magnet flux lines on the paper as shown by the arrangement of the filings.

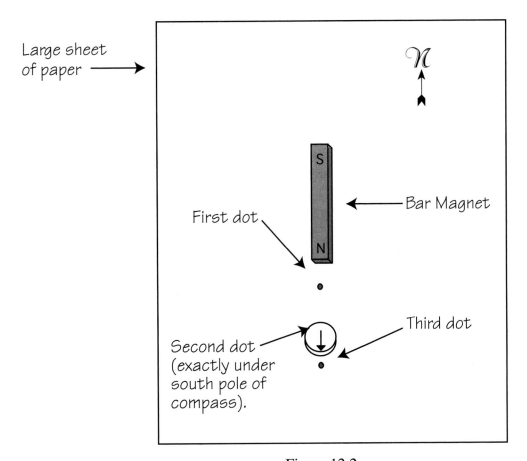

Figure 12.2

Results

1. How is using iron filings (a) similar and (b) different from using a magnetic compass to map a magnetic field?

2. In terms of a force, or torque, on a magnetic compass needle, what do the lines actually represent? Explain.

3. Do the lines ever cross each other at any point? Explain.

4. Where do the lines appear to be concentrated the most? What does this mean?

Experiment 13: Reflection and Refraction

Invitation to Inquiry

1. Boil water, then cool and freeze it in various shapes of lenses. Boiling the water drives the air out of solution, making clear ice. Use the ice lenses to ignite paper by focusing sunlight on a paper.

2. Make a prism by fastening three glass microscope slides together to form a triangle. Seal the edges and end with cellophane tape, fill it with water, then seal the other end. You can also seal the two ends by pushing them into a block of wax. Use the prism to experiment with sunlight and strong sources of light.

3. Fill a small aquarium with water that has been made murky by mixing in chalk dust, talcum powder, or some other insoluble substance that will make a beam of light visible. Experiment with aiming a strong beam of light through the side of the aquarium and up to the inner surface, directing the beam at increasing angles.

Background

The travel of light is often represented by a **light ray**, a line that is drawn to represent the straight-line movement of light. The line represents an imaginary thin beam of light that can be used to illustrate the laws of reflection and refraction, the topics of this laboratory investigation.

A light ray travels in a straight line from a source until it encounters some object. What happens next depends on several factors, including the nature of the material making up the object, the smoothness of the surface, and the angle at which the light ray strikes the surface. If the surface is perfectly smooth, rays of light undergo **reflection**. If the material is transparent, on the other hand, the light ray may be transmitted through the material. In these cases the light ray appears to become bent, undergoing a change in the direction of travel at the boundary between two transparent materials (such as air and water). The change of direction of a light ray at the boundary is called **refraction**.

Light rays traveling from a source, before they are reflected or refracted, are called *incident rays*. If an incident ray undergoes reflection, it is called a *reflected ray*. Likewise, an incident ray that undergoes refraction is called a *refracted ray*. In either case, a line perpendicular to the surface, at the point where the incident ray strikes, is called the *normal*. The angle between an incident ray and the normal is called the *angle of incidence*. The angle between a reflected ray and the normal is called the *angle of reflection*. The angle between a refracted ray and the normal is called the *angle of refraction*. These terms are descriptive of their meaning, but in each case you will need to remember that the angle is measured from a line perpendicular to the surface, the **normal**.

Procedure

Part A: Reflection of Light

1. Using a ruler, draw a straight line across a sheet of plain (unlined) white paper. Place the paper on a smooth piece of cardboard that has been cut from a box. Label the line with a B at one end and B´ at the other end (B is for boundary).

2. Attach a small, flat mirror to a block of wood as shown in figure 13.1. Place the mirror and block combination on the paper with the back of the mirror (the reflecting surface) on line BB´.

3. Stick a pin straight up and down into the paper about 10 cm from the mirror and slightly to the right side as shown in figure 13.1. On the left side, carefully align the edge of a ruler with the reflected image as shown in the illustration. Then firmly hold the ruler and draw a pencil line along this edge. Move the mirror and extend this line to the mirror boundary line BB´. Label the point of reflection with the letter P.

4. Place a protractor on line BB´ and mark a point 90° from the line. From this point, use the ruler to draw a dashed normal (NP). Complete your ray diagram by using the ruler to draw a line from the point of reflection (P) to the source of the light ray at the pin (I). Place arrows on line IP and line PR to show which way the light ray moved.

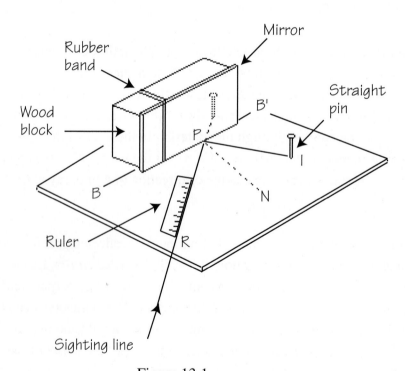

Figure 13.1

5. Use the protractor to measure the angle of incidence and the angle of reflection. Record these angles in Data Table 13.1 under Trial 1.

6. Place the mirror with its back edge again on line BB´ and conduct a second and third trial at different sighting angles. Record these measurements in Data Table 13.1 on page 112.

Part B: Refraction of Light

1. Place a clean sheet of white (unlined) paper on the cardboard. Place a glass plate approximately 5 cm square flat on the center of the paper. Use a pencil to outline the glass plate, then move the plate aside.

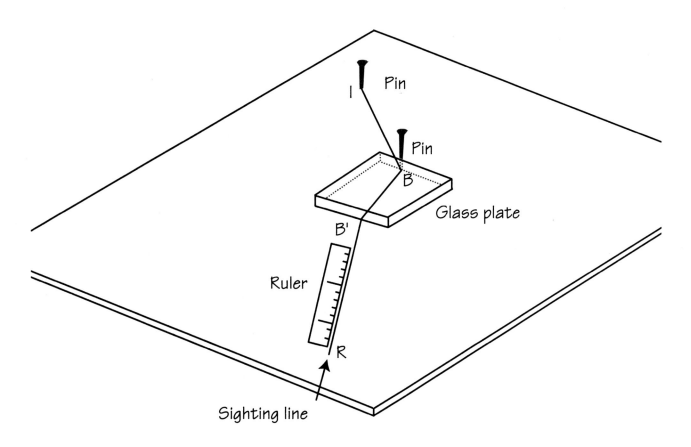

Figure 13.2

2. Use a ruler to draw a straight line from the upper left edge of the plate outline, making an angle of about 60° to the edge. Label this line IB as shown in figure 13.2. Place one upright pin at point B immediately outside the plate outline and a second upright pin at point I. Return the glass plate to the outline.

109

3. Bring the cardboard, paper, and glass plate near the edge of the table so you can sight through the glass plate toward the two pins. Position a ruler so that one edge aligns with the two pins as shown in figure 13.2. Draw a line along the ruler and label the line B´R. Move the glass plate aside for a second time.

4. Draw a line from B to B´, showing the path of the light ray through the glass. Overall, the path of the light ray is from IB to BB´ to B´R, showing that the light ray was bent twice.

5. Draw normals to the surface of the glass at B and B´. Show the angle of incidence and the angle of refraction with curved arrows at both boundaries.

Results

1. Describe any pattern you found in the data between the angle of incidence and the angle of reflection.

2. Describe what happens to a light ray as it travels (a) from air into glass and (b) from glass into air.

3. Assuming that light travels faster through air than it does through glass, make a generalized statement about what happens to a light ray with respect to the normal as it moves from a faster speed in one material to a slower speed in another.

4. What rules or generalizations do your findings suggest about reflection? How much more data would be required to make this a valid generalization?

5. What rules or generalizations do your findings suggest about refraction? How much more data would be required to make this a valid generalization?

6. Was the purpose of this lab accomplished? Why or why not? (Your answer to this question should be reasonable and make sense, showing thoughtful analysis and careful, thorough thinking.)

Trial	Angle of Incidence	Angle of Reflection
Data Table 13.1 Reflection of Light		
1	_____	_____
2	_____	_____
3	_____	_____

Experiment 14: Physical And Chemical Change

Invitation to Inquiry

In many communities the recycling of aluminum, paper, and plastics is started by first segregating items made of these materials from the rest of the trash. A major problem in recycling plastics is the many different types of plastics that exist, all with different chemical and physical properties. Some of these materials are more desirable for recycling than others, so they must be sorted. One way of sorting plastics is to read the code that might be stamped on the bottom. Here are some letter and number codes from some common plastic items:

The number code usually appears inside the recycling arrow logo. 2 (HDPE) milk jugs, bleach, and detergent bottles; 1 (PETE) soft-drink bottles; 5 (PP) ketchup bottles and yogurt cups; 6 (PS) transparent plastic drinking cups and CD boxes; and 4 (LDPE) plastic squeeze bottles. Can you find a way to separate a mixture of pieces of plastic from each of these 5 groups by taking advantage of the differences in chemical or physical properties? Consider cutting pieces of plastic from each of the group and finding important properties that could be used in a separation scheme.

Background

Matter undergoes many changes, and most of the common, everyday changes are either physical or chemical. A **physical change** is one that does not alter the identifying properties of a substance. It can be a change in form, state, or energy level, but no permanent change occurs in the properties of the substance. A piece of paper, torn into two parts, for example, still has the properties of the original paper, and no new substance has been formed. Thus tearing a piece of paper into two parts is a physical change. Evaporation, condensation, melting, freezing, and dissolving often produce physical changes.

A **chemical change** is one that produces new substances with new properties. A piece of paper that burns produces gases and ash that have different properties than the original paper, so this is an example of a chemical change. Heat, light, and electricity often produce chemical changes. Chemical changes occur as (1) atoms combine to form new compounds, (2) compounds break down into simpler substances, and (3) compounds react with other compounds or elements to form new substances. In this experiment you will determine if certain changes in matter are physical or chemical changes.

Procedure

1. Dissolve about 2 g of sodium chloride (ordinary table salt) in 50 mL of water in a clean graduated cylinder. Observe (a) if the water level changes when the salt is added and (b) the taste of the solution. **CAUTION:** Taste chemicals only when directed to do so. Record your observations in Data Table 14.1.

2. Continue dissolving 2 g samples of sodium chloride until 10 g are dissolved. Record your observations about the water level and the taste after each of the five additions.

3. Place about 5 mL of the solution in an evaporating dish and heat slowly until a dry solid remains. When the solid has cooled, carefully taste the solid and record your observation in the data table.

4. Carefully examine a 5-cm length of nichrome wire, noting the color, luster, flexibility, and other properties you can observe. Hold the wire with tongs and heat it strongly in the flame of a burner until it glows brightly. After the wire cools, again examine the wire, comparing the properties before and after heating. Record your observations.

5. Examine a 5-cm length of magnesium ribbon, noting the color, luster, flexibility, and other properties you can observe. Hold the ribbon with tongs and ignite the end in the flame of a burner. **CAUTION**: Do not look directly at the magnesium while it is burning. Compare the properties of the ash with the properties of the original magnesium ribbon, recording your observations in the data table.

6. Pour about 5 mL of silver nitrate solution in a small test tube. Add several drops of dilute hydrochloric acid, gently swirling the mixture with the addition of each drop. Record your observations in the data table. Filter the solid that forms (the precipitate) using small amounts of water as necessary to flush all the precipitate from the test tube. (Figure 14.1 shows how to

Step 1: Fold along diameter.   Step 2: Fold over a second time.

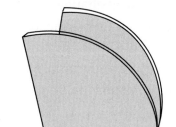

Step 3: Open one fold to make cone.   Step 4: Place cone in funnel.

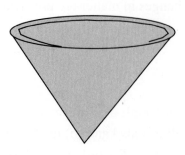

Figure 14.1

fold a filter paper, and figure 14.2 shows how to set up apparatus to filter a liquid.) Discard the filtered liquid (the filtrate) and place the filtered precipitate in direct sunlight. Record your observations after the precipitate has been in the sunlight for two or three minutes.

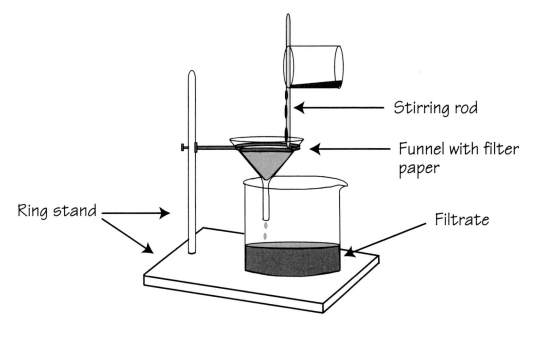

Figure 14.2

7. Pour about 125 mL of copper(II) chloride solution in a beaker. Observe if the beaker feels cool or warm to touch. Obtain a piece of aluminum foil approximately 2 cm by 15 cm and form it into a loose coil. Drop the aluminum foil coil into the copper(II) chloride solution. Observe the results, if any, and cautiously touch the beaker after about five minutes. Record your observations in the data table.

Results

1. What kind of change occurs when sodium chloride dissolves? Give an explanation for your answer.

2. What changes occurred with the water level, if any, as more and more sodium chloride was added to the solution? Provide an explanation for your observation.

3. Compare the evidence of a new substance being formed before and after heating (a) a nichrome wire and (b) a magnesium ribbon.

4. What evidence indicates if new substances with new properties were or were not formed when hydrochloric acid was added to a silver nitrate solution?

5. Did the precipitate exposed to sunlight undergo either a chemical or physical change? Give an explanation for your answer.

6. What changes did you observe when the aluminum was added to the copper(II) chloride solution? (Hint: There were more than five changes.)

7. Which of the changes in this experiment were physical changes? Give reasons for your conclusion.

8. Which of the changes in this experiment were chemical changes? Give reasons for your conclusion.

9. Was the purpose of this lab accomplished? Why or why not? (Your answer to this question should be reasonable and make sense, showing thoughtful analysis and careful, thorough thinking.)

Data Table 14.1	Physical and Chemical Changes

Action	Observations
Dissolving sodium chloride	
Evaporating sodium chloride solution	
Heating nichrome wire	
Heating magnesium ribbon	
Silver nitrate and hydrochloric acid mixture	
Precipitate in sunlight	
Aluminum foil in copper(II) chloride solution	

Experiment 15: Conductivity of Solutions

Invitation to Inquiry

Use some small gauge wire, a flashlight battery, and a flashlight bulb and holder to set up a portable conduction tester. Test a variety of materials to find which are conductors and which are nonconductors. Generalize what categories of materials are conductors and what categories seem to be nonconductors.

Background

Ionic compounds that dissolve in water do so by ions being pulled from the crystal lattice to form a solution of free ions. When free ions (charged particles) are present in a solution, the solution is a good conductor of electricity. Many **covalent compounds** that dissolve in water form molecular solutions of noncharged particles. Such a solution will not conduct an electric current. Some covalent compounds are pulled apart into ions; that is, the compound is *ionized* in a water solution. Strong ionization results in a solution that is a good conductor of electricity. Partial ionization results in a solution that is a poor conductor of electricity. The amount of current that flows through such a solution is roughly related to the amount of ionization. In this experiment you will use a conductivity apparatus (figure 15.1) to compare the conductivity of different solutions.

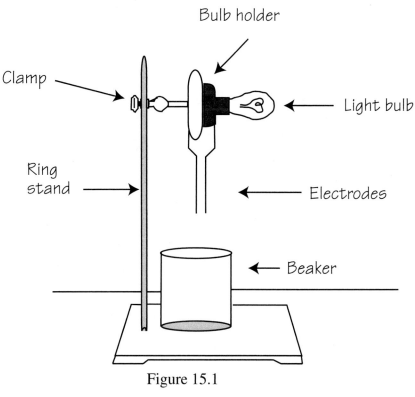

Figure 15.1

CAUTION: The solutions in this experiment are dilute and relatively safe provided that you exercise care in their use. Acids and bases, however, can irritate the skin and damage clothing and other materials. If you spill an acid or base on your skin or clothing, wash it off immediately with plenty of water and inform your instructor.

Procedure

1. Set up the apparatus as shown in figure 15.1. With the lamp *unplugged*, attach a patch cord with alligator clips to the two electrodes. Plug in the lamp to make sure the apparatus works, then unplug the lamp. Remove the patch cord and set it aside.

2. Test each substance listed in Data Table 15.1 using the conductivity apparatus with the following procedure:

   For solids: With the lamp *unplugged*, lower the electrodes until both are touching the solid. Plug in the lamp and record if it glows brightly, dimly, faintly, or not at all. *Unplug* the lamp and remove the electrodes from the solid. Wipe the electrodes with a clean cloth before proceeding.

   For solutions: With the lamp *unplugged*, lower the electrodes to the same depth in each solution. Plug in the lamp and record if the bulb glows brightly, dimly, faintly, or not at all. *Unplug* the lamp and remove the electrodes from the solution. Wash the electrodes by immersing them in distilled water before proceeding to the next solution test.

3. Your instructor might specify other solids or solutions to be tested which should be added to the list in Data Table 15.1.

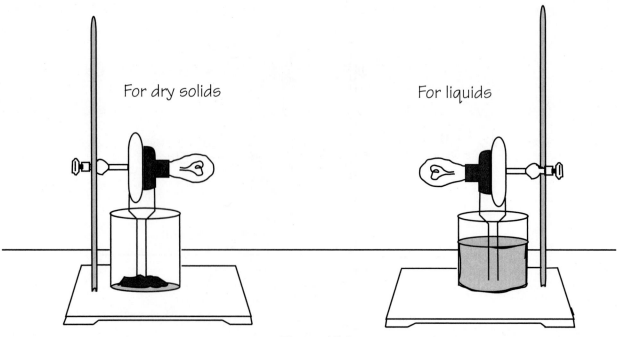

Figure 15.2

Results

1. Compare your observations of the conductivity of dry sodium chloride and a solution of sodium chloride. What conclusions can you make about these observations?

2. Compare your observations of the conductivity of distilled water and tap water. Offer an explanation for this comparison.

3. Compare the conductivity of hydrochloric acid and vinegar, which is also an acid solution. If the amount of conductivity is an indication of the degree of ionization of an acid solution, what would you predict about the degree of ionization of these two solutions?

4. What solutions were good conductors of electricity? What solutions were nonconductors of electricity? Describe any patterns that you can find in this summary.

5. In general, what kind of compounds conduct electricity when they are in solution? Explain.

6. In general, what kind of compounds do *not* conduct electricity when they are in solution? Is this always true? Explain.

7. Was the purpose of this lab accomplished? Why or why not? (Your answer to this question should be reasonable and make sense, showing thoughtful analysis and careful, thorough thinking.)

Data Table 15.1	Conductivity Tests of Solids and Solutions
Substance	Observations
Dry solid sodium chloride	
Distilled water	
Tap water	
Sodium chloride solution	
Hydrochloric acid	
Vinegar	
Ethyl alcohol	
Sodium hydroxide solution	
Glycerin	
Sugar solution	

Experiment 16: Metal Replacement Reactions

Invitation to Inquiry

Make a plan for using the fewest number of steps possible in an experimental study to place six metals (Mg, Zn, Fe, Sn, Cu, and Al) in an activity series from highest to lowest. Describe the procedure you would use for determining the position of a Ni salt in this series.

Background

Based on what happens to the reactants and products, there are basically four types of chemical reactions: (1) combination, (2) decomposition, (3) replacement, and (4) ion exchange reactions. This experiment is concerned with **metal replacement reactions**. A metal replacement reaction occurs when a more active metal is added to a solution of the salt of a less active metal. In generalized form the reaction is $XY + A \rightarrow AY + X$, where A represents the more active metal and XY represents the salt of a less active metal.

A metal replacement reaction occurs because some metals have a stronger electron holding ability than other metals. Metals that have the least ability to hold on to their electrons are the most chemically active. An activity series ranks the most chemically active metals at the top and the least chemically active at the bottom. This means that the activity series is upside down with respect to electron holding ability. Metals at the top of the activity series have the least ability to hold on to their electrons and those at the bottom have the greatest ability to hold on to their electrons.

A metal replacement reaction occurs when a more active metal (with less electron holding ability) is added to a solution containing the ions of a less active metal (with greater electron holding ability). The more active metal loses electrons to the less active metal, so they trade places; that is, the active metal loses electrons to form metallic ions in solution, and the less active metal gains electrons and comes out of solution as a solid metal. In this experiment you will study the chemical reaction of three metals and rank them according to their electron holding ability.

CAUTION: Soluble lead compounds are poisonous. Silver nitrate solutions will stain the skin. Avoid contact with both of these metal salt solutions. Wash thoroughly with soap and water if contact is suspected. Wash your hands after completing the experiment.

Procedure

1. Obtain three clean, dry test tubes and a test tube rack. Add 10 mL of silver nitrate solution to each test tube.

2. Obtain three metal strips each of copper, zinc, and lead. Sandpaper all nine strips until they are clean and bright. Make a right-angle bend near the end of each metal strip. A 10-cm length of thin copper wire with one end formed into a hook can be used to fish the bend of a metal strip from a test tube. Use the copper wire to remove the metal strips from their solution for inspection as needed.

3. Place a strip of each metal into a test tube with the silver nitrate solution. Observe the metals for 10 minutes or so, looking for evidence of a replacement reaction. Carefully touch each test tube at the outside bottom to observe any temperature changes. If a black deposit appears on a metal strip, observe for an additional 5 minutes or longer. In Data Table 16.1 record your observations of any changes in the solutions and any changes on the metal strips.

4. Repeat procedure steps 1 through 3 with fresh pieces of metal and 10 mL of copper nitrate solution. Observe the metals for 10 minutes or more, again looking for evidence of a replacement reaction.

5. Repeat procedure steps 1 through 3 again, this time with fresh pieces of metal and 10 mL of lead nitrate solution. Again observe the metals for 10 minutes or longer and record your observations in the data table.

Results

1. According to Data Table 16.1, which metals had the (a) greatest number of reactions; (b) middle number of reactions; (c) least number of reactions?

2. Are your findings in question 1 consistent with the activity series of metals? Explain.

3. Were any changes observed in the color of the solutions? Offer an explanation for this observation.

4. Complete the following equations, writing "no reaction" if none occurred:

$Cu + AgNO_3 \rightarrow$

$Zn + AgNO_3 \rightarrow$

$Pb + AgNO_3 \rightarrow$

$Cu + Cu(NO_3)_2 \rightarrow$

$Zn + Cu(NO_3)_2 \rightarrow$

$Pb + Cu(NO_3)_2 \rightarrow$

$Cu + Pb(NO_3)_2 \rightarrow$

$$Zn + Pb(NO_3)_2 \rightarrow$$

$$Pb + Pb(NO_3)_2 \rightarrow$$

5. Write a general rule describing when metal replacement reactions occur and when they do not occur.

6. Was the purpose of this lab accomplished? Why or why not? (Your answer to this question should be reasonable and make sense, showing thoughtful analysis and careful, thorough thinking.)

Data Table 16.1	Metal Replacement Reactions		
	Zinc metal	Copper metal	Lead metal
Silver nitrate solution			
Copper nitrate solution			
Lead nitrate solution			

Experiment 17: Identifying Salts

Invitation to Inquiry

Foods naturally contain enzymes, biochemical compounds that originate in plants and animals. Cooking changes enzymes, and the purpose of blanching or steaming food shortly before freezing is intended to destroy enzymes. One of the enzymes in foods is a catalyst that will speed the decomposition of hydrogen. You can design a home experiment to use hydrogen peroxide (3% solution) to test fresh, blanched, and cooked crushed food (or juices from the foods) for the catalyst enzymes. If you can find a way to quantify the measurements, perhaps you can come up with specific recommendations about how hot the food should be heated and for how long.

Temperature is one of the more important factors that influence the rate of a chemical reaction. You can use a "light stick" or "light tube" to study how temperature can influence a chemical reaction. Light sticks and tubes are devices that glow in the dark and have become very popular on July 4th, Halloween, and other times when people might be outside after sunset. They work from a chemical reaction that is similar to the chemical reaction that produces light in a firefly. Design a home experiment that uses light sticks to find out the effect of temperature on the brightness of light and how long the device will continue providing light. Perhaps you will be able to show by experimental evidence that use at a particular temperature produces the most light for the longest period of time.

Background

A **salt** is any ionic compound except those with hydroxide (OH^-) or oxide (O^{-2}) ions. Table salt, which is sodium chloride ($NaCl$), is but one example from the large group of ionic compounds known as salts. Simple salts consist of a metallic ion (such as Na^+) in an ionic crystalline structure with nonmetallic ions (such as Cl^-). When a salt is dissolved in water the ion crystal structure is separated into a solution of metallic and nonmetallic ions.

The name of a salt provides a clue about the ions present in a solution. Lithium chloride ($LiCl$) becomes Li^+ and Cl^- ions when dissolved in water. Likewise, calcium iodide (CaI_2) becomes a solution of Ca^{+2} and I^- ions, and iron(II) carbonate [$Fe_2(CO_3)_3$] becomes a solution of Fe^{+3} and CO_3^{-2} ions. As the water of a salt solution is evaporated, the salt ions again form an ionic crystal structure. Knowing the ions present in a given salt solution will therefore identify the salt in the solution. In this experiment, you will use a flame test to identify metallic ions present in a salt solution. Each metal ion gives a characteristic color to a burner flame. The nonmetallic ions will be identified by chemical tests.

CAUTION: Acids will damage human tissue and other materials. Silver nitrate will stain the skin. Some of the chemicals are poisonous when swallowed. Be sure to wash thoroughly if any chemicals are spilled and inform your instructor. Wash your hands thoroughly after the experiment.

Procedure

Part A: Flame Tests

1. Obtain seven test tubes and a test tube rack. Wash the test tubes and rinse them thoroughly with distilled water. Cleanliness is very important throughout this experiment.

2. Pour about 25 mL of dilute hydrochloric acid into one test tube and label it in a rack. Pour about 5 mL of each of the six nitrate solutions into test tubes and label each in the rack. Set up a burner and adjust it for an inner blue cone and no yellow color.

3. Clean a platinum or nichrome wire by dipping it into dilute hydrochloric acid then holding it in the pale blue part of the flame. Repeat the cleaning procedure until the wire gives no color to the flame. The flame test you are about to do is a very sensitive test, and even trace contamination of sodium from your fingers will give the characteristic color of the sodium ion (a yellow coloration).

4. Dip the clean wire into one of the solutions and place just the tip of the wire into the light blue burner flame. Record your observations in Data Table 17.1.

5. Clean the wire, then repeat the flame test with each of the solutions. Record the results obtained for each in Data Table 17.1. Use two thicknesses of cobalt glass to view the potassium and sodium flames. The sodium ion gives a color that masks the color given by the potassium ion. The cobalt glass filters out the color produced by any sodium ions.

Part B: Chemical Tests

1. Obtain four clean test tubes and a test tube rack. Pour about 5 mL of each solution of sodium chloride, potassium sulfate, calcium carbonate, and one of the nitrate solutions from the flame test into separate test tubes. Label each in the test tube rack.

2. **Chloride ion test**: To 5 mL of sodium chloride solution, add 5 drops of silver nitrate solution. Mix the solutions well, then describe the results in Data Table 17.2.

3. **Sulfate ion test**: To 5 mL of potassium sulfate solution, add 10 drops of barium chloride solution. Mix well, then describe the results in Data Table 17.2.

4. **Carbonate test**: To 5 mL of calcium carbonate solution, add several drops of dilute hydrochloric acid. Describe the results in Data Table 17.2.

5. **Nitrate ion test**: To 5 mL of a nitrate solution, add 10 mL of iron(II) sulfate solution. Use a long, thin pipette to slowly add 2 mL of concentrated sulfuric acid so it runs down the inside of the tube without mixing. Observe the interface between the acid and the solution, recording the results in Data Table 17.2.

Part C: Unknown Solutions

1. Wash test tubes of all salt solutions and rinse thoroughly with distilled water.

2. Your instructor will supply unknown solutions, each containing a single metallic and a single nonmetallic ion. Using the procedures and findings from Part A and Part B, identify the ions in the unknown solutions. Record your findings in Data Table 17.3, describing your test results as very positive, positive, trace, negative, or unsure.

Data Table 17.1	Flame Tests for Metallic Ions
Metallic ion	Results
Sodium	
Lithium	
Strontium	
Calcium	
Barium	
Potassium	

Data Table 17.2	Chemical Tests for Nonmetallic Ions
Nonmetallic ion	Results
Chloride	
Sulfate	
Carbonate	
Nitrate	

Data Table 17.3	Unknown Solutions		
Unknown	Metallic Ion Tests	Nonmetallic Ion Tests	Compound
1	Sodium: Lithium: Strontium: Calcium: Barium: Potassium:	Chloride: Sulfate: Carbonate: Nitrate:	Name: _____ Formula: _____
2	Sodium: Lithium: Strontium: Calcium: Barium: Potassium:	Chloride: Sulfate: Carbonate: Nitrate:	Name: _____ Formula: _____
3	Sodium: Lithium: Strontium: Calcium: Barium: Potassium:	Chloride: Sulfate: Carbonate: Nitrate:	Name: _____ Formula: _____

Experiment 18: Natural Water

Invitation to Inquiry

Washing soda, borax, or trisodium phosphate are often added to laundry products to soften the wash water. Add a small amount of each to 5 mL of tap water, then measure the hardness with the soap-drop method described in this experiment. Is one of the softening chemicals more effective than the others?

Is there a practical way to obtain pure water from a hard water source? Design, use, and evaluate an apparatus for purifying water by boiling, freezing, filtering, precipitation of dissolved minerals, or by any means you can imagine. Evaluate your technique in terms of usefulness and effectiveness.

Background

Water is sometimes referred to as the "universal solvent" because so many gases, liquids, and solids readily dissolve in it. In addition, the solubility of carbon dioxide in water produces an acid (H_2CO_3) that contributes to the dissolution of normally insoluble carbonate, phosphate, and sulfite minerals. Thus natural water can contain a significant amount of dissolved mineral matter as well as suspended solids. This occurs naturally from weathering and erosion of the earth's surface by rainwater and by erosion of stream beds.

Natural water that contains dissolved mineral salts in the form of calcium ions or magnesium ions is called **hard water**. It is called hard water because the metal ions react with soap, making it hard to make soap lather. Instead of producing suds, soap in hard water produces an insoluble, sticky precipitate. This requires a greater amount of soap to produce suds since all the metal ions must be precipitated out of solution before the soap will lather. The precipitate also produces a bathtub ring, collects on clothes, and results in other undesirable effects. When boiled, hard water can produce a solid deposit that restricts water flow in water heaters and pipes.

Water hardness is usually measured in parts per million (ppm), with hard water identified as any concentration of calcium or magnesium ions greater than about 75 ppm. In this experiment you will learn a method of analyzing water hardness as well as analyzing the type of hardness present in your water supply.

Procedure

Part A: Suspended Solids

1. Natural water usually contains suspended solids that are removed by (a) natural settling, (b) addition of a gelatinous precipitate, and (c) filtering through sand. Compare these methods by obtaining four beakers each with 100 mL of muddy water.

2. Allow beaker #1 of muddy water to stand undisturbed as you complete the rest of this section. This will provide a comparison of natural settling to the other methods of removing suspended solids.

3. Make a sand filter by placing a loose plug of glass wool in a funnel, then covering it with several centimeters of clean sand. Pour beaker #2 of muddy water through the filter and collect the filtrate in a clean beaker.

4. In beakers #3 and #4, produce the gelatinous precipitate of aluminum hydroxide by adding to each 10 mL of alum solution then 4 drops of ammonium hydroxide while stirring. Allow both suspensions to settle.

5. Prepare a second sand filter as in procedure step 3, then filter one of the settled beakers of aluminum hydroxide suspension from procedure step 4. Collect this filtrate in a clean beaker and save it for part C of this experiment.

6. Compare the appearance of the muddy water from each procedure step in Data Table 18.1.

Part B: A Test for Hard Water

1. The calcium and magnesium ions of hard water react with soap, forming an insoluble precipitate. Soap will form suds only after the ions have been removed, so the amount of soap needed to produce suds is an indication of the amount of water hardness.

2. Measure 5 mL of the calcium chloride solution and pour it into a test tube. This solution has been prepared to have a water hardness of 100 ppm.

3. Add one drop of soap solution to the calcium chloride solution and insert a clean stopper. Shake the test tube vigorously, then check the surface for soap suds. If none are visible, or if visible suds do not persist for at least one minute, add a second drop and shake the test tube vigorously again. Continue adding one drop at a time and shaking, keeping track of the number of drops added until a layer of suds persists for at least a minute. Record the number of drops required for permanent suds in Data Table 18.2.

4. Repeat procedure Part B step 3 with a clean test tube and 5 mL of distilled water. A certain number of soap drops will be required to make a permanent suds layer because of the mechanism by which suds are produced. In Data Table 18.2, record the number of soap drops required for this mechanism to occur in distilled water.

5. Complete the calculations required in Data Table 18.2 to find the hardness equivalent to one drop of the soap solution.

Part C: Water Hardness

1. Repeat procedure step B-3 with 5 mL of the flocculated and filtered water saved from part A of this experiment. Record the number of drops of soap required in Data Table 18.3 and calculate the hardness of the water.

2. Repeat the soap-drop test with 5 mL of ordinary cold tap or well water. Record the number of soap drops required in Data Table 18.3 and calculate the water hardness.

3. Water hardness caused by calcium or magnesium bicarbonate can be removed by boiling, so it is called temporary hardness. The hardness remaining after boiling is permanent hardness. Temporary hardness is therefore the difference between total hardness and permanent hardness.

4. Pour 25 mL of cold tap water or well water into a small beaker. Place the beaker on a wire screen on a ring stand with a clean watch glass on top of the beaker to prevent water loss by splattering. Gently heat the water to boiling, continuing a slow boil for 5 or 6 minutes.

5. Cool the water to room temperature, then pour the water into a graduated cylinder. Add distilled water as necessary to bring the volume back to 25 mL. Pour 5 mL into a test tube. Use the soap-drop method to determine the permanent hardness. Calculate the temporary hardness. Record all data in Data Table 18.3.

Results

1. Compare the appearance of muddy water after natural settling and the other means of removing suspended solids. Which method produced the best apparent purity? Is there any impurity that the best method does not remove?

2. Explain how a soap solution was standardized to measure water hardness.

3. What is hard water?

4. Explain how temporary and permanent hardness can be measured.

5. What is the white solid that forms on water outlets in hard water areas? How could you test the substance to confirm your answer?

6. Was the purpose of this lab accomplished? Why or why not? (Your answer to this question should show thoughtful analysis and careful, thorough thinking.)

Data Table 18.1	Removing Suspended Solids
Action	Result
1. Initial appearance of muddy water	
2. After natural settling	
3. After sand filtration only	
4. After aluminium hydroxide flocculation and settling	
5. After aluminum hydroxide flocculation, settling, and sand filtering	

Data Table 18.2 Hardness Equivalent of Soap Solution	
1. Drops of soap solution required for calcium chloride solution (100 ppm)	_____ drops
2. Drops of soap solution for distilled water -------------------------	_____ drops
3. Subtract sudsing mechanism (row 1 – row 2) -------------------	_____ drops
4. Hardness equivalent (100 ppm ÷ row 3) -------------------------	_____ ppm/drop

Data Table 18.3　　Permanent, Temporary, and Total Hardness of Tap Water

1. Drops of soap solution required for filtered water --------------- _____ drops

2. Water hardness (row 1 × ppm/drop from Data Table 18.2) ------ _____ ppm

3. Drops of soap solution required for tap water -------------------- _____ drops

4. Total water hardness of tap water --------------------------------- _____ ppm

5. Drops of soap solution required after boiling --------------------- _____ drops

6. Permanent hardness of tap water ---------------------------------- _____ ppm

7. Temporary hardness of tap water (row 4 – row 6) --------------- _____ ppm

Experiment 19: Measurement of pH

Invitation to Inquiry

There are many common things that can be used as acid and base indicators. For example, from foods you could try boiled purple cabbage juice, grape juice, blackberry juice, ordinary tea, and others, and from the office supply store certain construction papers might show color changes. To test foods, try soaking strips of filter paper in juices or solutions then allowing it to dry completely. Materials such as construction paper can be tested directly. Plan tests to find what color each indicator will appear when exposed to solutions of acids or bases, and test other materials, too. Show how your collections of indicators will identify the pH from a wide range of possibilities.

Think of some way to measure the "strength" of an acid or a base without using an indicator. Taking proper precautions, do experiments to compare several methods of measuring strength.

Background

Acids and bases are classes of chemical compounds, and each class has certain properties in common. Acids have characteristic properties of reacting with active metals to release hydrogen gas, neutralizing bases to form a salt, and changing the color of certain substances such as litmus. Bases, on the other hand, have characteristic properties of converting plant and animal tissues to soluble substances, neutralizing acids to form a salt, and reversing the color changes that were caused by acids.

Some chemical compounds are acidic, such as vinegar, and other compounds are basic, such as lye, but the terms acidic and basic are very general and inexact. The **pH scale** is a concise and quantitative way of expressing the strength of an acid or a base. On this scale a neutral solution (neither acidic nor basic properties) has a pH of 7.0. Acid solutions have pH values below 7, and smaller numbers mean greater acidic properties. Basic solutions have pH values above 7, and larger numbers mean greater basic properties. The pH scale is logarithmic, so a pH of 2 is 10 times more acidic than a pH of 3. Likewise, a pH of 8 is 10 times 10 or 100 times more basic than a pH of 6.

Many plant extracts and synthetic dyes change colors when mixed with acids or bases. Such substances that change color in the presence of acids or bases can be used as an **acid-base indicator**. Litmus, for example, is an acid-base indicator made from a dye extracted from certain species of lichens. The dye is applied to paper strips which turn red in acidic solutions and blue in basic solutions. This is only a "ball park" indicator, however, since it only indicates on which side of pH 7 a solution lies. Other indicators are available that can be used to estimate the pH to about half a pH unit—for example, a pH of 3.5. There are also electrical instruments that measure pH by measuring the conductivity of a solution. In this experiment you will determine the characteristics of several commercial indicators then test unknown solutions for acid and base properties.

CAUTION: Acids and bases can damage skin and other materials. If spilled on your person or clothing, flush with running water for several minutes and inform the instructor. Dilute any tabletop spills with water, then use a sponge to wipe up the spill. Be sure to rinse the sponge with plenty of water.

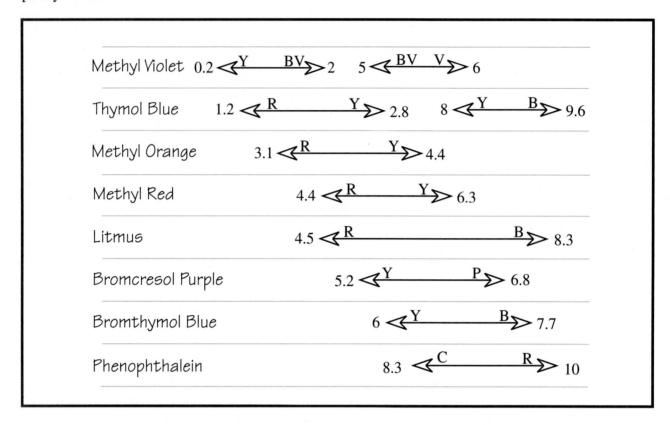

Figure 19.1 pH Ranges and Color Changes for Selected Indicators

Procedure

1. Obtain five clean test tubes and a test tube rack. Pour 5 mL of hydrochloric acid into each. Test the hydrochloric acid with litmus by placing a strip of red litmus paper and a strip of blue litmus paper in a clean watch glass. Dip a clean, dry stirring rod into the solution in one of the test tubes, then transfer a drop to each litmus paper. Record your observations in Data Table 19.1. Save the litmus paper and watch glass for further testing.

2. Add two drops of each indicator solution to a test tube of hydrochloric acid, recording your observations in Data Table 19.1. Include pH values or ranges where possible.

3. Place a strip of universal indicator paper on the watch glass. Use the stirring rod to transfer a drop of hydrochloric acid to the indicator paper. The universal indicator paper container will have a color standard chart with a corresponding pH. Record the color and pH number in Data Table 19.1. If a color falls between two of the standards, estimate the pH to a fraction between the corresponding pH values.

4. Repeat procedure steps 1 through 3 using sulfuric acid, acetic acid, sodium hydroxide, barium hydroxide, and dilute household ammonia solutions. Record the results for the acid solutions in Data Table 19.1 and the results for the base solutions in Data Table 19.2. When all of the solutions have been tested, take a few minutes to analyze your findings. Compare the pH ranges of the indicators in table 19.1 with the pH as shown by the universal indicator.

5. Obtain an unknown solution and use the different indicator tests to determine if the solution is an acid or a base and to determine the pH of the solution. Record your test results and findings in Data Table 19.3.

Results

1. A popular noncarbonated beverage turns thymol blue to yellow and methyl orange to red. What is the approximate pH of this soft drink?

2. What is the approximate pH of a beer that turns methyl orange to a yellow color?

3. How would the color changes of a mixture of methyl orange, methyl red, bromthymol blue, and phenolphthalein compare with those of the universal indicator?

4. Would you use a solution of phenolphthalein to find the strength of an acid? Explain.

147

5. The flowering shrub hydrangea produces blue flowers in a certain soil but produces pink flowers if lime is added to the soil. Propose an explanation for this observation. Explain how you could do tests on the soil to check your explanation.

6. Was the purpose of this lab accomplished? Why or why not? (Your answer to this question should show thoughtful analysis and careful, thorough thinking.)

Data Table 19.1 — Results of Indicator Tests for Acid Solutions

Indicator	Hydrochloric Acid	Sulfuric Acid	Acetic Acid
Red Litmus			
Blue Litmus			
Bromthymol Blue			
Methyl Orange			
Methyl Red			
Phenolphthalein			
Universal Indicator Color			
Universal Indicator pH			

Data Table 19.2	Results of Indicator Tests for Base Solutions		
Indicator	Sodium Hydroxide	Barium Hydroxide	Dilute Ammonia
Red Litmus			
Blue Litmus			
Bromthymol Blue			
Methyl Orange			
Methyl Red			
Phenolphthalein			
Universal Indicator Color			
Universal Indicator pH			

Data Table 19.3	Test Results for Unknown Solution
Indicator	Results
Red Litmus	
Blue Litmus	
Bromthymol Blue	
Methyl Orange	
Methyl Red	
Phenolphthalein	
Universal Indicator Color	

pH of unknown solution: _____

Is the unknown solution an acid or a base? _____

Experiment 20: Amount of Water Vapor in the Air

Invitation to Inquiry

Cobalt chloride is often used to test for the presence of water since it undergoes a reversible color change when exposed to moisture or humidity. For example, cobalt chloride is sometimes included with silica gel pellets to indicate when the gel has absorbed moisture. Experiment with cobalt chloride dried on filter paper strips. Find out if the color change is sensitive enough to water vapor and temperature to be used as an indicator of the relative humidity.

Background

The amount of water vapor in the air is referred to generally as **humidity**. A measurement of the amount of water vapor in the air at a particular time is called the **absolute humidity**. At room temperature, for example, humid air might contain 15 grams of water vapor in each cubic meter of air. At the same temperature air of low humidity might have an absolute humidity of only 2 grams per cubic meter. Absolute humidity can range from near zero up to a maximum that is determined by the temperature at a particular time, as shown in figure 20.1. Since the temperature of the water vapor present in the air is the same as the temperature of the air, the maximum absolute humidity is usually said to be determined by the air temperature. What this really means is that the maximum absolute humidity is determined by the temperature of the water vapor; that is, the average kinetic energy of the water vapor.

The relationship between the *actual* absolute humidity at a particular temperature and the *maximum* absolute humidity that can occur at that temperature is called the **relative humidity**. Relative humidity is a ratio between (1) the amount of water vapor in the air and (2) the amount of water vapor needed to saturate the air at that temperature. The relationship is

$$\frac{\text{actual absolute humidity at present temperature}}{\text{maximum absolute humidity at present temperature}} \times 100\% = \text{relative humidity}.$$

For example, suppose a measurement of the water vapor in the air at 10° C (50° F) finds an absolute humidity of 5 g/m³. According to figure 20.1, the maximum amount of water vapor that can be in the air when the temperature is 10° C is about 10 g/m³. The relative humidity is then

$$\frac{5 \, g/m^3}{10 \, g/m^3} \times 100\% = 50\%.$$

If the absolute humidity were 10 g/m³, then the air would have all the water vapor it could hold and the relative humidity would be 100%. A relative humidity of 100% means that the air is saturated at the present temperature.

Procedure

Part A: Maximum Amount of Water Vapor

1. Measure the present air temperature in your laboratory room and record it in Data Table 20.1. Use the graph of maximum absolute humidity in figure 20.1 to estimate the *maximum* amount of water vapor a cubic meter of air can hold at this temperature. Since one gram of water has an approximate volume of one milliliter, find the maximum amount of water in liters per cubic meter that can be in the room at the present temperature. Record this maximum, in grams/cubic meter and in liters/cubic meter, in Data Table 20.1.

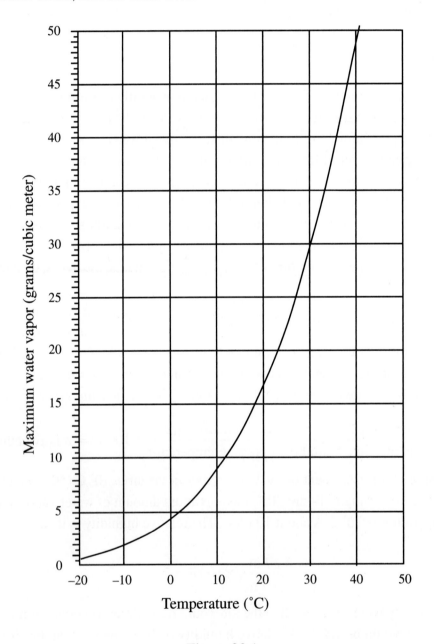

Figure 20.1

2. Measure the length, width, and height of the laboratory room. Record these measurements in Data Table 20.1, then calculate the volume of the room in cubic meters. Record all data and calculations in the data table.

3. Calculate the maximum amount of water vapor the air in the laboratory room can hold at the present temperature. This is found by multiplying the volume of the room (in cubic meters) by the maximum amount of water vapor that could be in the room at the present temperature (in liters per cubic meter). Record all data and calculations in Data Table 20.1.

Part B: Actual Amount of Water Vapor

Evaporation occurs at a rate that is inversely proportional to the relative humidity, ranging from a maximum rate when the air is driest to no net evaporation when the air is saturated. Since evaporation is a cooling process, it is possible to use a thermometer to measure humidity. An instrument called a **psychrometer** has two thermometers, one of which has a damp cloth wick around its bulb end. As air moves past the two thermometer bulbs, the ordinary thermometer (the dry bulb) will measure the present air temperature. Water will evaporate from the wet wick (the wet bulb) until an equilibrium is reached between water vapor leaving the wick and water vapor returning to the wick from the air. Since evaporation lowers the temperature, the depression of the temperature of the wet-bulb thermometer is an indirect measure of the water vapor present in the air. The relative humidity can be determined by obtaining the dry- and wet-bulb temperature readings and referring to a relative humidity chart such as the one found inside the back cover. If the humidity is 100%, no net evaporation will take place from the wet bulb, and both wet- and dry-bulb temperatures will be the same. The lower the humidity, the greater the difference in the temperature reading of the two thermometers.

Relative humidity is a ratio between the actual absolute humidity at a given temperature and the maximum absolute humidity that can occur at that temperature. Knowing the maximum absolute humidity and the relative humidity, you can find the amount of water vapor in the air at the present temperature.

1. Wet the cotton wick on the wet bulb of a sling psychrometer. Whirl the thermometers in the air until the wet-bulb thermometer registers its lowest reading. Record the wet-bulb and dry-bulb temperatures, then use this data to find the relative humidity from the relative humidity chart inside the back cover. Record all data and calculations in the data table.

2. Multiply the humidity as a fraction times the maximum amount of water vapor that could be in your laboratory room at the present temperature. Record the amount of water vapor present in the room in the data table.

Results

1. What is the actual amount of water vapor present in the laboratory room air at the present temperature (in g/m^3)? What is the maximum amount of water vapor that *could* be present?

2. Can the absolute humidity of the air in the room be increased? Explain.

3. Can the relative humidity of the air in the room be increased without adding more water vapor to the air? Explain.

4. Suppose the room air has all the water vapor it will hold at 25° C and the air is cooled to 15° C. Considering the area of the floor from your measurements, how deep a layer of water will condense from the air?

5. Was the purpose of this lab accomplished? Why or why not? (Your answer to this question should show thoughtful analysis and careful, thorough thinking.)

Data Table 20.1	Amount of Water Vapor in the Laboratory Room	
1. Present temperature of laboratory room air	_____	°C
2. Maximum absolute humidity at present temperature		
In grams per cubic meter	_____	g/m^3
In liters per cubic meter	_____	L/m^3
Room length	_____	m
Room width	_____	m
Room height	_____	m
3. Volume of laboratory room (length × width × height)	_____	m^3
4. Maximum amount of water vapor (row 2 × row 3)	_____	L
5. Dry-bulb reading	_____	°C
Wet-bulb reading	_____	°C
Difference in wet- and dry-bulb readings	_____	°C
Relative Humidity (from relative humidity chart)	_____	%
6. Amount of water vapor in room (row 5 as decimal × row 4)	_____	L

Experiment 21: Growing Crystals

Invitation to Inquiry

Investigate the crystal structure of some large samples of crystals. Consider, for example, measuring the angles between the faces, comparing how the crystal transmits light, and other measurements that would help you interpret the structure in terms of atoms or molecules. Does the structure provide any information about the conditions necessary for growing large and well-formed crystals? Or are the same conditions needed for all crystals? Are there any upper limits on the size of a crystal?

Background

A salt dissolves in water because the polar water molecules pull ions away from the crystal lattice of the salt. The oxygen end of a water molecule has a negative polar charge, and the hydrogen ends are positive. The oxygen ends of water molecules tend to orient themselves toward the positive ions on the outside of the crystal lattice as opposite charges attract one another. The hydrogen ends likewise orient themselves toward the negative ions of the lattice. If the attraction of water molecules is greater than the attraction between the ions in the lattice, the ions are pulled away and dissolution occurs. Saturation occurs as water molecules become "tied up" in their attraction for ions. Fewer water molecules means less attraction on the ionic solid with more solute ions being pulled back to the surface of the crystal lattice.

Each salt forms its own kind of crystal which is one way to identify salts as well as other crystalline substances such as minerals that occur in the earth's crust. In this investigation a warm, saturated solution is cooled, and this lowers the solubility. Rapid cooling—or rapid evaporation of water—results in many small crystals that can be seen with a hand lens. If the conditions are favorable, large, beautiful crystals can be "grown" from a saturated solution. This investigation provides a procedure for growing large, beautiful crystals and will require an extended time period.

CAUTION: Some salts can be poisonous if taken internally. Wash hands thoroughly after handling salts or solutions.

Procedure

1. Using Table 21.1, select a salt to produce a large crystal. Note that several different chemicals may be sold as "alum" by drug and grocery stores. If alum is selected and purchased from a drug or

grocery store, make sure it is potassium aluminum sulfate.

2. Using a ring stand and ring, wire screen, and large Pyrex beaker, heat 1 L of distilled water. When the water is hot, begin stirring as you slowly add the approximate amount listed for the salt selected from table 21.1. Stir while slowly adding the salt and continue stirring until no more will dissolve. Increase the solution temperature as necessary, stirring until you are confident that you have a saturated solution. If all the salt dissolves, continue adding small amounts until no more will dissolve.

3. Remove the burner flame, allowing the solution to cool slightly. Decant (pour the solution from any undissolved salts in the bottom) the saturated solution while still warm into a large glass jar. Thread a length of fishing leader through a hole punched in the center of the lid, then tape it securely to the lid. When the lid is on the jar the leader should be long enough to extend halfway into the solution. If the leader floats, use the stirring rod to push it below the surface of the liquid. Place the jar where it can remain undisturbed overnight.

4. Small crystals will form in the solution overnight. Carefully remove the lid and place the fishing leader across a paper towel. Remove all crystals from the leader, selecting the one largest crystal as your seed crystal. Epoxy this seed crystal to the end of the dry leader.

5. Pour the solution from the jar back into the large beaker, being sure that all crystals (except the seed crystal) go with the solution. Heat the solution and again prepare a saturated solution. When you are confident that you have a saturated solution, cool it to room temperature, then return the seed crystal to the solution. Adjust the fishing leader and tape it to the lid so the seed crystal is about centered in the solution. If your seed crystal dissolves, you did not have a saturated solution or the solution was not cooled to room temperature. If you make this error, return to procedure step 3.

6. The solution and seed crystal are now allowed to stand undisturbed at a constant temperature for several days. To produce large, perfect crystals it is absolutely necessary for the solution to be isolated from physical disturbances and at a constant temperature. In addition, any small crystals that form on the surface of the seed crystal during a growth period must be completely removed by breaking, scraping, and sanding.

7. The first growth period is complete when the crystal stops growing. When this occurs remove the crystal from the solution and provide more ions by repeating procedure steps 5 and 6. This process is repeated for more growth periods until the crystal reaches the desired size. Any imperfections that develop are the result of some physical or temperature disturbance.

Results

1. Explain why sudden temperature changes result in irregular growth.

2. Why is it necessary to remove other, smaller crystals from the solution?

3. Why is it necessary to re-saturate the solution for each growth period?

4. Why does each salt form its own kind of crystal?

5. Was the purpose of this lab accomplished? Why or why not? (Your answer to this question should show thoughtful analysis and careful, thorough thinking.)

Salt	Crystal Color	Approximate Amount Needed: More than. . .
Potassium aluminum sulfate (alum)	Colorless	225 g
Sodium chloride (table salt)	Colorless	425 g
Copper(II) sulfate (blue vitriol)	Blue	2.02 kg
Nickel sulfate (hydrate)	Blue or Green	3.53 or 4.83 kg
Cobalt(II) sulfate	Red	860 g
Cobalt(II) chloride (hydrate)	Red	1.2 kg
Potassium chromate	Yellow	825 g

Table 21.1   Salts, Crystal Colors, and Amounts Needed to Grow Crystals

Experiment 22: Properties of Common Minerals

Invitation to Inquiry

Build or obtain a tumbler that gives minerals a smooth, polished surface. Investigate and predict which locally available minerals would be best for polishing. Learn how to use the tumbler and prepare some local specimens for a display.

Background

The earth's crust is made up of rocks, which are solid aggregations of materials that have been cohesively brought together by rock-forming processes. The fundamental building blocks of rocks are **minerals**, which are naturally occurring, inorganic solid elements or chemical compounds with a crystalline structure. About 2,500 different minerals are known to exist, but only about twenty are common in the crust. Examples of these common minerals are quartz, calcite, and gypsum.

Each rock-forming mineral has its own set of physical properties because each mineral has (1) a chemical composition and (2) a crystal structure that is unlike any other mineral. The exact composition and crystal structure of an unknown mineral can be determined by using laboratory procedures. This type of analysis is necessary when the crystal structures are too small to be visible to the unaided eye. In this laboratory activity you will learn how to identify minerals by considering some identifying characteristics of large, well-developed mineral crystals.

Procedure

Part A: Developing a Table of Mineral Properties

Examine each mineral specimen in the mineral collection set, one at a time. Consider each of the following properties and record all observations in Data Table 22.1.

1. **Color.** The color of a mineral specimen is an obvious property, but the color of some minerals varies from one specimen to the next because of chemical impurities. Sometimes the property of color can be useful in identifying a mineral. Use combinations of words as necessary to describe mineral colors—for example, reddish-brown.

2. **Streak.** Streak is the color of a mineral when it is finely powdered, a property that is more consistent from specimen to specimen than color. Streak is observed by rubbing the corner of a specimen across an unglazed streak plate. Determine the streak color of each mineral and record your observations in Data Table 22.1. Note that a mineral harder than the streak plate will not

leave a streak. This is useful information, so record this observation if it occurs.

3. **Hardness.** Hardness is the resistance of a mineral to being scratched. The Mohs hardness scale is used as a basis for measuring hardness. The hardness test is done by trying to scratch a mineral with a mineral (or some object with an equivalent hardness) from the Mohs scale. If the mineral being investigated scratches a Mohs hardness scale mineral, the mineral in question is harder than the test mineral. If the mineral being investigated is scratched by the test mineral, the mineral in question is not as hard as the test mineral. If both minerals are scratched by each other, they have the same hardness. The Mohs hardness scale is given in table 22.1 along with the hardness of some common objects that can be used for hardness tests. A hardness of 1 on this scale is assigned to the softest mineral, and the hardest mineral has an assigned hardness of 10.

Table 22.1	Mohs Scale of Hardness for Minerals and Common Objects Tests	
Hardness Number	Mineral Example	Common Object Test
1	Talc	Can be scratched with fingernail.
2	Gypsum	Can be scratched with fingernail, but not easily.
3	Calcite	Can be scratched with copper penny.
4	Fluorite	Can be scratched with pocketknife. Will not scratch glass.
5	Apatite	Can be scratched with pocketknife, but not easily.
6	Feldspar	Can be scratched with edge of steel file. Cannot be scratched with pocketknife.
7	Quartz	Scratches glass.
8	Topaz	Scratches quartz.
9	Corundum (ruby or sapphire, for example)	Scratches all minerals but diamond.
10	Diamond	Scratches all minerals. Can be scratched only by another diamond.

4. **Cleavage.** Cleavage is the tendency of a mineral to break along smooth planes. Cleavage occurs in parallel planes in some minerals and in one or more directions in other minerals. Sometimes cleavage is perfect; other minerals may have indistinct cleavage.

5. **Fracture.** Minerals that do not have cleavage may show fracture, an irregular broken surface rather than the smooth planes of cleavage. Some minerals have a distinct way of fracturing, such as the conchoidal fracture of obsidian and quartz. Conchoidal fracture is the breaking along curved surfaces like a shell.

6. **Luster.** Luster describes the surface sheen, the way a mineral reflects light. Minerals that have the surface sheen of a metal are described as being metallic. Nonmetallic luster is described by such terms as pearly (like a pearl), vitreous (like glass), glossy, dull, and so forth.

7. **Density.** Density is a ratio of the mass of a mineral to its volume. Often the density of a mineral is expressed as specific gravity, which is a ratio of the mineral density to the density of water. To obtain an exact specific gravity, a mineral specimen must be pure and without cracks, bubbles, or substitutions of chemically similar elements.

8. **Other properties.** There are a few other properties that are specific for one or more minerals. These properties are determined by tests such as a reaction to acid, reaction to a magnet, and others. Some of the specific properties of certain minerals, such as the double image seen through a calcite crystal, are unique enough to identify an unknown mineral in an instant.

Part B: Unknown Minerals

Once the Table of Mineral Properties (Data Table 22.1) is complete, you can use it to identify an unknown mineral. The procedure is find out what the mineral is not. For example, suppose you start with the streak test and find the unknown mineral leaves a red streak. This test tells you the unknown mineral is not one of the minerals that leave some other color of streak. Then you look at the minerals that have a red streak and perform a second test. The second test eliminates still other possibilities. Eventually, you find what the unknown mineral is by finding out what it isn't.

Results: Was the purpose of this lab accomplished? Why or why not? (Your answer to this question should show thoughtful analysis and careful, thorough thinking.)

Mineral	Color	Luster	Streak
Calcite			
Magnetite			
Pyrite			
Hematite			
Limonite			
Quartz			
Fluorite			
Apatite			
Feldspar			
Talc			
Hornblende			
Halite			
Bauxite			
Biotite			
Muscovite			
Sphalerite			
Galena			
Chrysotile			
Gypsum			

Data Table 22.1 Table of Mineral Properties

Data Table 22.1	Table of Mineral Properties, Continued		
Mineral	Hardness	Cleavage/Fracture	Other
Calcite			
Magnetite			
Pyrite			
Hematite			
Limonite			
Quartz			
Fluorite			
Apatite			
Feldspar			
Talc			
Hornblende			
Halite			
Bauxite			
Biotite			
Muscovite			
Sphalerite			
Galena			
Chrysotile			
Gypsum			

Experiment 23: Density of Igneous Rocks

Invitation to Inquiry

Survey the use of rocks used in building construction in your community. Compare the type of rocks that are used for building interiors and those that are used for building exteriors. Are any trends apparent for buildings constructed in the past and those built in more recent times? If so, are there reasons (cost, shipping, other limitations) underlying a trend, or is it simply a matter of style?

Background

Igneous rocks are rocks that form from the cooling of a hot, molten mass of rock material. Igneous rocks, as other rocks, are made up of various combinations of minerals. Each mineral has its own temperature range at which it begins to crystallize, forming a solid material. Minerals that are rich in iron and magnesium tend to crystallize at high temperatures. Minerals that are rich in silicon and poor in iron and magnesium tend to crystallize at lower temperatures. Thus minerals rich in iron and magnesium crystallize first in a deep molten mass of rock material, sinking to the bottom. The minerals that crystallize later will become progressively richer in silicon as more and more iron and magnesium are removed from the melt.

Igneous rocks that are rich in silicon and poor in iron and magnesium are comparatively light in density and color. The most common igneous rock of this type is **granite**, which makes up most of the earth's continents. Igneous rocks that are rich in iron and magnesium are dark in color and have a relatively high density. The most common example of these dark-colored, more dense rocks is **basalt**, which makes up the ocean basins and much of the earth's interior. Basalt is also found on the earth's surface as a result of volcanic activity.

Procedure

1. Use a balance to find the mass of a basalt rock. Record the mass in Data Table 23.1. Tie a 20 cm length of nylon string around the rock so you can lift it with the string. Test your tying abilities to make sure you can lift the rock by lifting the string without the rock falling.

2. Place an overflow can on a ring stand, adjusted so the overflow spout is directly over a graduated cylinder.

3. Hold a finger over the overflow spout, then fill the can with water. Remove your finger from the spout, allowing the excess water to flow into the cylinder. Dump this water from the cylinder, then

place it back under the overflow spout.

4. Grasp the free end of the string tied around the basalt, then lower the rock completely beneath the water surface in the overflow can. The volume of water that flows into the graduated cylinder is the volume of the rock. Remembering that a volume of 1.0 mL is equivalent to a volume of 1.0 cm^3, record the volume of the rock in cm^3 in Data Table 23.1.

5. Calculate the mass density of the basalt and record the value in the data table.

6. Repeat procedure steps 1 through 5 with a sample of granite.

Results

1. In what ways do igneous rocks have different properties?

2. Explain the theoretical process or processes responsible for producing the different properties of igneous rocks.

3. According to the experimental evidence of this investigation, propose an explanation for the observation that the bulk of the earth's continents are granite and that basalt is mostly found in the earth's interior.

4. Was the purpose of this lab accomplished? Why or why not? (Your answer to this question should show thoughtful analysis and careful, thorough thinking.)

Data Table 23.1	Density of Igneous Rocks		
	Mass (g)	Volume (cm^3)	Density (g/cm^3)
Basalt	_____ g	_____ cm^3	_____ g/cm^3
Granite	_____ g	_____ cm^3	_____ g/cm^3

Experiment 24: Latitude and Longitude

Invitation to Inquiry

There are several ways to find your latitude by measurement. First, determine your latitude by measuring the angle of the North Star above the horizon. Second, determine your latitude by measuring the angle between a vertical stick and a line to the noonday sun on the spring equinox (March 21) or the autumnal equinox (September 23). For the North Star, consider making two measurements 12 hours apart and averaging the two. Why do these two different methods tell you your latitude? Is one more in "agreement" with the stated latitude for your location?

Background

The continuous rotation and revolution of the earth establish an objective way to determine directions and locations on the earth. If the earth were an unmoving sphere there would be no side, end, or point to provide a referent for directions and locations. The earth's rotation, however, defines an axis of rotation which serves as a reference point for determination of directions and locations on the entire surface. The reference point for a sphere is not as simple as on a flat, two-dimensional surface because a sphere does not have a top or side edge. The earth's axis provides the north-south reference point. The equator is a big circle around the earth that is exactly halfway between the two ends or poles of the rotational axis. An infinite number of circles are imagined to run around the earth parallel to the equator. The east- and west-running parallel circles are called **parallels**. Each parallel is the same distance between the equator and one of the poles all the way around the earth. The distance from the equator to a point on a parallel is called the **latitude** of that point. Latitude tells you how far north or south a point is from the equator by telling you on which parallel the point is located.

Since a parallel is a circle, a location of 40° N latitude could be anyplace on that circle around the earth. To identify a location you need another line, one that runs pole to pole and perpendicular to the parallels. North-south running arcs that intersect at both poles are called **meridians**. There is no naturally occurring, identifiable meridian that can be used as a point of reference such as the equator serves for parallels, so one is identified as the referent by international agreement. The reference meridian is the one that passes through the Greenwich Observatory near London, England and is called the **prime meridian**. The distance from the prime meridian east or west is called the **longitude**. The degrees of longitude of a point on a parallel are measured to the east or to the west from the prime meridian up to 180°.

Locations identified with degrees of latitude north or south of the equator and degrees of longitude east or west of the prime meridian are more precisely identified by dividing each degree of latitude into subdivisions of 60 minutes (60') per degree and each minute into 60 seconds (60"). In this investigation you will do a hands-on activity that will help you understand how latitude and longitude are used to locate places on the earth's surface.

Procedure

1. Obtain a lump of clay about the size of your fist. Knead the clay until it is soft and pliable, then form it into a smooth ball for a model of the earth.

2. Obtain a sharpened pencil. Hold the clay ball in one hand and use a twisting motion to force the pencil all the way through the ball of clay. Reform the clay into a smooth ball as necessary. This pencil represents the earth's axis, an imaginary line about which the earth rotates. Hold the clay ball so the eraser end of the pencil is at the top. The eraser end of the pencil represents the North Pole, and the sharpened end represents the South Pole. With the North Pole at the top, the earth turns so the part facing you moves from left to right. Hold the clay ball with the pencil end at the top and turn the ball like this to visualize the turning earth.

3. The earth's axis provides a north-south reference point. The equator is a circle around the earth that is exactly halfway between the two poles. Use the end of a toothpick to make a line in the clay representing the equator.

4. Hold the clay in one hand with the pencil between two fingers. Carefully remove the pencil from the clay with a back and forth twisting motion. Reform the clay into a smooth ball if necessary, being careful not to destroy the equator line. Use a knife to slowly and carefully cut halfway through the equator. Make a second cut down through the North Pole to cut away one-fourth of the ball as shown in figure 24.1. Set the cut-away section aside for now.

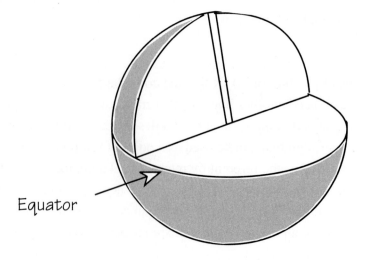

Equator

Figure 24.1

5. Place a protractor on the clay ball where the section was removed. As shown in figure 24.2, the 0° of the protractor should be on the equator and the 90° line should be on the axis (the center of where the pencil was). You may have to force the protractor slightly into the clay so the 0° line is on the equator. Directly behind the protractor, stick toothpicks into the surface of the clay ball at 20°, 40°, and 60° above the equator on both sides. Remove the protractor from the clay and return the cut-away section to make a whole ball again.

6. Use the end of a toothpick to make parallels at 20°, 40°, and 60° north of the equator, then remove the six toothpicks. Recall that parallels are east and west running circles that are the same distance from the equator all the way around the earth (thus the name parallels). The distance from the equator to a point on a parallel is called the **latitude** of that point. Latitude tells you how far north or south a point is from the equator. There can be 90 parallels between the equator and the North Pole, so a latitude can range from 0° North (on the equator) up to 90° North (at the North Pole).

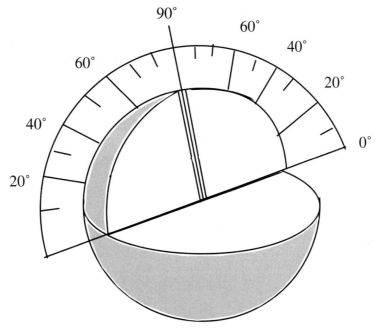

Figure 24.2

7. Since a parallel is a circle that runs all the way around the earth, a second line is needed to identify a specific location. This second line runs from pole to pole and is called a meridian. To see how meridians identify specific locations, again remove the cut section from the ball of clay. This time place the protractor flat on the equator as shown in figure 24.3, with the 90° line perpendicular to the axis. Stick toothpicks directly below the protractor at 0°, 60°, 120°, and 180°, then remove the protractor and return the cut-away section to make a whole ball again. Use a toothpick to draw lines in the clay that run from one pole, through the toothpicks, then through the other pole. By agreement, the 0° line runs through Greenwich near London, England, and this meridian is called the prime meridian. The distance east or west of the prime meridian is called **longitude**. If you move right from the prime meridian you are moving east from 0° all the way to 180° East. If you move left from the prime meridian you are moving west from 0° all the way to 180° West.

8. Use a map or a globe to locate some city that is on or near a whole number latitude and longitude. New Orleans, Louisiana, for example, has a latitude of about 30° N of the equator. It has a longitude of about 90° W of the prime meridian. The location of New Orleans on the earth is therefore described as 30° N, 90° W. Locate this position on your clay model of the earth and insert a toothpick. Compare your model to those of your classmates.

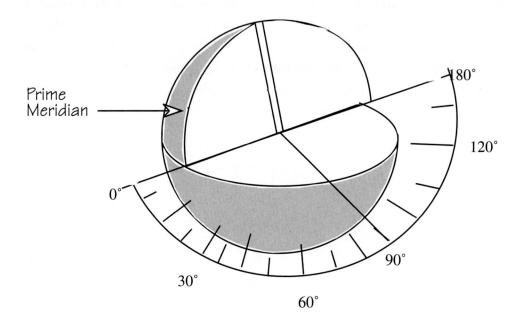

Figure 24.3

Results

1. What information does the latitude of a location tell you?

2. What information does the longitude of a location tell you?

3. According to a map or a globe, what is the approximate latitude and longitude of the place where you live?

4. Explain how minutes and seconds are used to identify a location more precisely.

5. Was the purpose of this lab accomplished? Why or why not? (Your answer to this question should show thoughtful analysis and careful, thorough thinking.)

Experiment 25: Telescopes

Invitation to Inquiry

Design an experiment to study the effect of the diameter of a lens on the image formed. Do the experiment.

Use plane (flat), concave, and convex mirrors to find when you can see

- an enlarged image.
- a reduced image.
- an image of the same size.
- an image that appears upright.
- an image that appears inverted.
- an image that appears upright, then inverted after some adjustment.

What generalizations can you make to inform someone how to make the various images with mirrors?

Background

A convex lens can be used as a "burning glass" by moving the lens back and forth until the sunlight is focused into a small bright spot—an image of the sun. This image is hot enough to scorch the paper, perhaps setting it on fire. The lens is moved back and forth to refract the parallel rays of light from the sun the point where the image is formed on the paper. The place where the image forms is called the **focal point** of the lens. The distance from the focal point to the lens is called the **focal length** (f). The focal length of a lens is determined by its index of refraction and the shape of the lens. The focal length is an indication of the refracting ability or strength of a lens. A lens with a short focal length is considered to be a stronger lens than one with a longer focal length.

There are three important measurements that are used to describe how lenses work as optical devices. These are (1) the *focal length* (f); (2) the *image distance* (d_i), the distance from the lens that an image is formed; and (3) the *object distance* (d_o), the distance from the object being imaged to the lens. The relationship between these measurements is given in the **lens equation**, which is

$$\frac{1}{f} = \frac{1}{d_o} + \frac{1}{d_i}.$$

The magnification produced by a lens is defined as the ratio of the height of the image (h_i) to the height of the object (h_o). This is also equal to the ratio of the image distance (d_i) to the object distance (d_o), or

$$\text{Magnification} \;=\; \frac{d_i}{d_o}.$$

In this investigation you will compare measuring the focal length of a lens directly with using the lens equation to calculate the focal length of a convex lens. Magnification of a lens will also be investigated by comparing direct measurement of magnification with theoretical magnification as calculated from the focal lengths of two lenses.

Procedure

1. Measure the focal length of a lens directly. First, place a lens in a lens holder and secure it at the 50 cm mark on a meter stick. Second, point the meter stick at some distant objects, such as a tree or a house about a block away. Move a cardboard screen in a holder back and forth until you obtain a sharp image on the screen. Finally, measure the distance between the sharp image and the lens, which is the focal length (f) of the lens. Record the focal length in Data Table 25.1 on page 186.

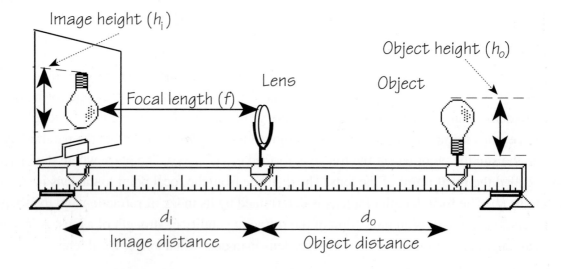

Figure 25.1

2. Find the focal length of the lens used in procedure step 1 by use of the lens equation. Set up a meter stick in holders, a screen in a holder, and a luminous object in a holder as shown in figure 25.1. The room should be darkened, then place the screen at the focal length distance found in procedure step 1. With the screen and object fixed in place, slowly move the lens along the meter stick to obtain the sharpest image possible. Note that the object and image should lie on a straight

182

line along and perpendicular to the principal axis of the lens. Measure and record in Data Table 25.1 the object distance (d_o) and the image distance (d_i) to the nearest 1 mm. Measure and record to the nearest 0.5 mm the height of the object (h_o) and height of the image (h_i). Record other observations here.

3. Place the lens *less than one focal length* from the bulb, then move the screen back and forth to see if you can obtain an image on the screen. Look through the lens at the bulb. Record your observations here.

4. Repeat procedure steps 1 and 2 for three more lenses. Record all data and results of calculations in Data Table 25.1. Select one of the lenses with a short focal length (for eyepiece lens) and one with a longer focal length (for objective lens) for use in the next procedure step.

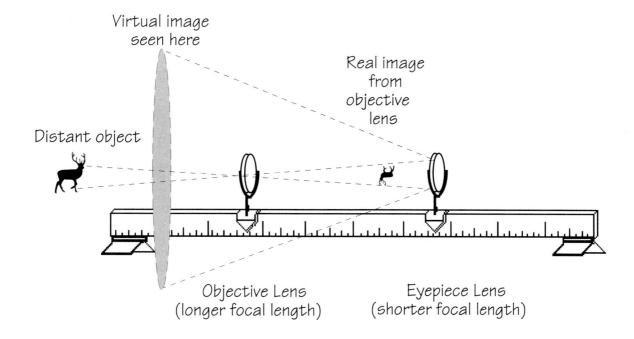

Figure 25.2

5. Make a telescope by mounting short and longer focal length lenses on a meter stick as shown in figure 25.2. Focus the telescope by adjusting the shorter focal length lens until it magnifies the image from the longer focal length lens. Calculate the *theoretical* magnification of your telescope by dividing the focal length of the objective lens by the focal length of the eyepiece lens. Record the theoretical magnification in Data Table 25.2.

6. Work with a partner to *determine experimentally* the magnification of your telescope. Focus your telescope on some object in front of your partner, who should hold a meter stick next to the object. Look at the object normally with one eye and look at the object through the telescope with the other eye. Direct your partner to position a pointer on the meter stick to indicate the apparent size of the enlarged image. Record the height of the image and the height of the object in Data Table 25.2, then calculate the magnification.

Results

1. What relationships did you find between d_o, d_i, and f?

2. What relationships did you find between d_o, d_i, h_i, and h_o?

3. Discuss the advantages, disadvantages, and possible sources of error involved in the two ways of finding the focal length of lenses.

4. Discuss the advantages, disadvantages, and possible sources of error involved in the two ways of finding the magnification of lenses.

5. Was the purpose of this lab accomplished? Why or why not? (Your answer to this question should show thoughtful analysis and careful, thorough thinking.)

Lens	Focal length measured directly (f)	Object distance (d_o)	Image distance (d_i)	Focal length from lens equation (f)	Object size (h_o)	Image size (h_i)	Magnification
1	_____	_____	_____	_____	_____	_____	_____
2	_____	_____	_____	_____	_____	_____	_____
3	_____	_____	_____	_____	_____	_____	_____
4	_____	_____	_____	_____	_____	_____	_____

Data Table 25.1   Lens Focal Length and Magnification

Data Table 25.2	Lens Magnification
Objective lens focal length	_____
Eyepiece lens focal length	_____
Theoretical magnification (Focal length of objective ÷ focal length of eyepiece)	_____
Object height (h_o)	_____
Image height (h_i)	_____
Experimental magnification ($h_i \div h_o$)	_____

Experiment 26: Celestial Coordinates

Background

To an observer unencumbered by scientific models, the night sky appears to be an inverted bowl resting on a flat plane. The observer appears to be located at the center of the bowl. Sprinkled over the inner surface of the bowl are the stars arranged in fixed, identifiable patterns that do not change noticeably from day to day, year to year, or even century to century.

Among what ancient observers called the "fixed stars" seven objects move: the sun, the moon, and five apparently star-like objects called planets. Other objects that moved in the sky but were transitory, such as meteors and comets, were considered by the ancient observers, Aristotle in particular, to be atmospheric phenomena and were not considered in modeling the heavens.

The motion of the sun was readily apparent. The general direction of its rising is synonymous with what we call the east. The direction of its setting is the west. Each day the sun rises to its highest altitude above the horizon when it is due south. If we imagine a line extending from the north point on the horizon passing overhead and continuing to the south point on the horizon, then the sun reaches its highest altitude when it crosses this line called the **meridian**. When the sun, or any other object in the sky, crosses the meridian, the object is said to **transit**.

During the course of a single night, the stars wheel around a fixed point as if the celestial bowl on which they are seemingly attached were spinning on an axis passing through this point and the observer's position. On a time exposure photograph made with a camera pointed toward this fixed point, each star traces out an arc of a perfect circle. Careful observation shows that the bowl rotates from east to west at a uniform rate, completing one revolution in just under 24 hours (23h56m4.s091). That is, the time interval between successive meridian transits of a particular star, say, is a bit shorter than the average time between successive meridian transits of the sun, 24 hours exactly. (The word "average" is used because the time interval between successive solar transits varies throughout the year.)

Since the sun rate and the star rate are close but not quite the same, it must be that the sun moves slowly with respect to the background stars, and in fact, the sun's path among the stars can be delineated. By noticing the sun's position among the stars at sunrise or sunset each day, we can plot its path among the stars. This path turns out to be a circle and is called the **ecliptic**, and the sun creeps from *west to east* at a nearly uniform rate, taking 365 days and almost 6 hours to complete a single circuit. Where is this circle on the celestial sphere? In order to answer this question, we will establish a coordinate system *on the sky* so that we can specify locations of points and circles. We will produce an analog of the system of latitude and longitude used to locate positions on the earth's surface.

Procedure

Part A: The Celestial Pole and Equator

Let us call the fixed point about which the sphere of stars turns the celestial pole. The direction on the ground toward this fixed point is called *north* if we are in the northern hemisphere of the earth. If we are in the southern hemisphere, a different fixed point is apparent, and the direction toward the point is *south*. In the northern sky a bright star, Polaris, is near the north celestial pole. There is no bright star near the south celestial pole.

On the surface of the earth, the line that is everywhere 90° away from the pole is called the equator. It is a circle bisecting the earth halfway between each pole. In the sky, the line (circle) on the celestial sphere that is everywhere 90° from the celestial pole is called the celestial equator. Figure 26.1 shows the geometry of the celestial sphere relative to that of the earth. The angle α formed by the observer, the center of the earth, and a point on the equator is defined to be the observer's latitude on the earth.

1. Prove that the angle α, the observer's latitude, is in fact the same as the angle (alpha) of the north celestial pole over the northern horizon.

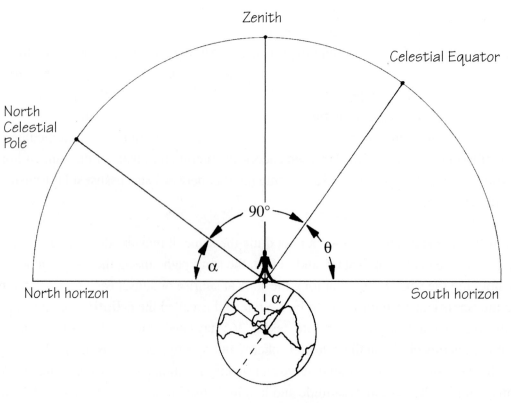

Figure 26.1

If the celestial pole is an angle α above the northern horizon, then we can establish the location of the celestial equator. From figure 26.1, the angle along the meridian that the equator makes over the southern horizon is $\theta = 180° - 90° - \alpha$. So the celestial equator starts in the east, climbs up to an angle of $90° - \alpha$ over the southern horizon, and then drops down to the west point of the horizon.

2. If the celestial pole makes an angle of 40° over the northern horizon, what angle over the southern horizon does the equator make?

3. Suppose you are at Quito, Ecuador (latitude = 0°). Where is the celestial equator?

The ecliptic is a circle on the celestial sphere that is tipped at an angle of 23 1/2° to the celestial equator. Figure 26.2 illustrates the relationship between the celestial equator and the ecliptic. Since during the course of a year the sun traces out the path of the ecliptic, it appears at different places with respect to the celestial equator. At one time, the sun is maximally above (north of) the celestial equator, sometimes right on the equator, and at another time maximally below (south of) the equator. Figure 26.3 shows two different orientations of the sun with respect to the celestial equator. The figures are drawn for the case where the north celestial pole lies at an angle of 40° over the north point of the horizon (e.g., latitude = 40° north). The figures are also drawn for noon when the sun is on the meridian.

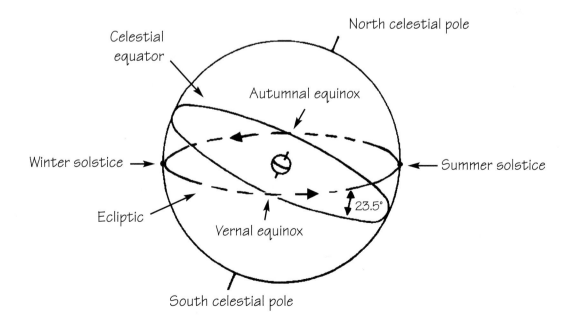

Figure 26.2 The ecliptic (dashed line) intersects the celestial equator at two points call equinoxes. The solstices mark the most northerly and southerly points.

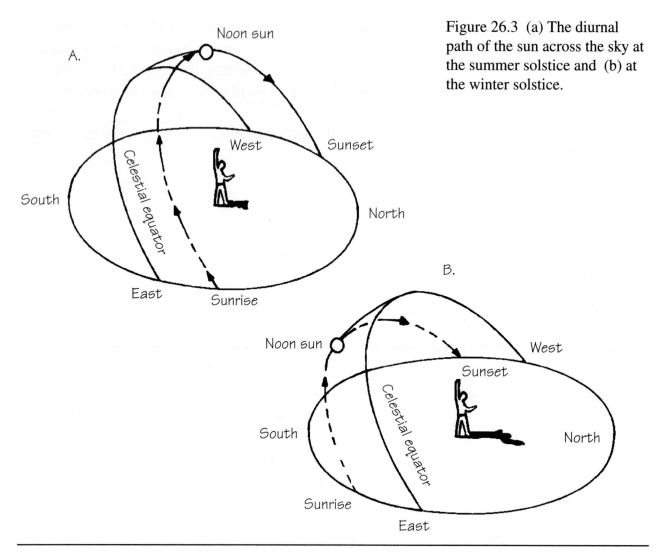

Figure 26.3 (a) The diurnal path of the sun across the sky at the summer solstice and (b) at the winter solstice.

Figure 26.4 The declination of an object on the celestial sphere is its angular distance measured in degrees north (+) or south (−) of the celestial equator.

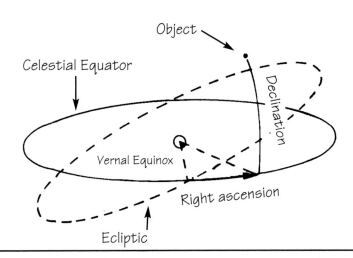

Figure 26.5 The right ascension of an object on the celestial sphere.

Part B: Celestial Coordinates

We can now use the points and circles we have discussed thus far to establish a coordinate system to map the heavens. In analogy with the latitude-longitude system used to specify the location of a point on the earth's surface, we will need two coordinates, called **declination** and **right ascension** to locate a particular point in the sky. The declination of a star is its angular distance measured in degrees between the celestial equator and the point or celestial object. The object's declination is positive (+) if it is north of the celestial equator and negative (–) if it is south. Subdivisions are measured in the usual minutes and seconds of arc. Figure 26.4 illustrates this coordinate. Points on the celestial equator have a declination of 0°; the north celestial pole has dec = +90°; the south pole has dec = –90°.

The right ascension coordinate is a bit more complicated. The right ascension of an object is the angular distance measured along the celestial equator between the vernal equinox and the point on the equator that intersects an hour circle passing through the object (see figure 26.5). An hour circle is a great circle that passes through both celestial poles as well as through the object itself.

Here is the peculiar aspect of the right ascension coordinate: it is measured in **hours**, with subdivisions of minutes and seconds of **time**. Instead of 360° all the way around the celestial equator, right ascension has 24h; that is 24h is equivalent to 360°. Right ascension is measured eastward from the vernal equinox.

4. How many degrees of arc are equivalent to 1^h of right ascension?

5. How many minutes of right ascension are equivalent to 1° of arc?

6. How many minutes of arc are equivalent to 1^m of right ascension?

Part C: The Celestial Globe

A celestial globe is a representation of the celestial sphere, the stars or other objects in the heavens, together with the coordinate system we have been discussing. Remember, the celestial sphere has you, the observer, at its center, so everything in the celestial globe is plotted from this perspective. As you look at the globe, you will have to visualize that you are at the center of the globe looking out.

7. Locate on the globe the (1) north and south celestial poles, (2) the celestial equator, (3) the ecliptic, (4) the two equinoxes where the ecliptic and equator intersect, and (5) the solstices.

8. Which stars are located at the following coordinates?

Right Ascension	Declination	Star Name
$6^h\ 43^m$	$-16°\ 39'$	
$6^h\ 22^m$	$-52°\ 40'$	
$18^h\ 35^m$	$+38°\ 44'$	
$5^h\ 12^m$	$-8°\ 15'$	
$7^h\ 37^m$	$+5°\ 21'$	
$14^h\ 36^m$	$-60°\ 38'$	

9. Which constellation is located at the approximate positions indicted?

Right Ascension	Declination	Constellation
11^h	$+50°$	
19^h	$-25°$	
3^h	$+20°$	
1^h	$+60°$	
13^h	$-50°$	
7^h	$-40°$	

The celestial globe is pivoted about the celestial poles and is held in place by a vertical circular ring of metal that is graduated in degrees. A large horizontal ring serves as a representation of the horizon. Adjust the vertical ring so that the celestial pole is at the proper altitude over the horizon for your latitude. (You may need to ask the laboratory instructor for your latitude to the nearest degree.)

As you rotate the globe you will notice that the celestial equator always maintains a single orientation in the sky; it comes up out of the east, reaches a maximum altitude when it crosses the meridian (vertical metal circle), and sinks in the west.

Part D: Relative Amounts of Daylight and Darkness

On the ecliptic locate the position of the sun on June 21. Place a small piece of 3M self-stick paper to mark the location. *Do not write on the globe or make any permanent markings.* Note the sun's declination for this day and record it in table 26.1. Now rotate the globe until the sun is in its rising position on the eastern horizon.

10. Does the sun rise exactly in the east on this day?

Rotate the globe until the sun sets and note where on the horizon the sun disappears below the horizon. Bring the sun back to the rising position and by counting how many right ascension hour circles lie between the rising and setting points for the sun, estimate to the nearest 15 minutes the amount of time the sun is above the horizon on June 21. Enter your estimate in table 26.1. By subtracting this number from 24^h, calculate the length of darkness for this date.

Repeat this for December 21 and March 21.

Readjust the position of the celestial pole to correspond to an observer at Altengaard, Norway (latitude = 70° north). Fill in table 26.2 as you did table 26.1.

Readjust the globe one more time to correspond to an observer in Quito, Ecuador (latitude = 0°). Fill in table 26.3.

11. *Calculate* the altitude of the sun (in degrees) over the southern horizon at noon on June 21 for an observer at latitude +40° N. Check your answer using the celestial globe.

12. Repeat procedure Part D step 11 for December 21.

195

Table 26.1 Relative Amounts of Daylight and Darkness for Your Latitude at Different Times of the Year

Date	Sec. of Sun	Number of of hours of Daylight	Number of Hours of Darkness
June 21			
December 21			
March 21			

Table 26.2 Relative Amounts of Daylight and Darkness for Altengaard, Norway (Latitude = 70° N) at Different Times of the Year

Date	Sec. of Sun	Number of of hours of Daylight	Number of Hours of Darkness
June 21			
December 21			
March 21			

Table 26.3   Relative Amounts of Daylight and Darkness for Quito, Equador (Latitude = 0°) at Different Times of the Year

Date	Sec. of Sun	Number of of hours of Daylight	Number of Hours of Darkness
June 21			
December 21			
March 21			

13. Eratosthenes, who lived in the second century B.C., was a Renaissance man before his time. He was an astronomer, a geographer, a historian, a mathematician, and a poet. With such a diverse background, it is not surprising that he was the director of the Great Library of Alexandria. In one of the scrolls at the library, Eratosthenes read that at noon on June 21 in the southern frontier outpost of Syene, Egypt, near the first cataract of the Nile, obelisks cast no shadows. As noon approached, the shadows of temple columns grew shorter until at noon they were gone. Reportedly, a reflection of the sun could then be seen at the bottom of a well. On June 21 at Syene, the sun was directly overhead. Use the celestial globe to determine the latitude of Syene.

14. Readjust the celestial globe for your latitude. What range of declinations can stars have so that they never set, that is, are always above the horizon?

Your instructor will provide you with the coordinates of Mercury and Venus for the day of this lab session. Put marked pieces of 3M self-stick paper at the corresponding locations on the celestial globe. Locate the sun for today and place a piece of paper at the sun's location on the globe.

15. When is Mercury visible in a dark sky? (circle one)

    just after sunset

    just before sunrise

16. When is Venus visible in a dark sky?

    just after sunset

    just before sunrise

17. Estimate how long each is visible in a dark sky.

    Mercury _____

    Venus_____

Experiment 27: Motions of the Sun

Background

(Note: Experiment 26, "Celestial Coordinates" is a prerequisite for this exercise). The sun rises above the eastern horizon, traces an arched path across the sky, and then sinks below the western horizon. Midway between sunrise and sunset the sun climbs to its highest altitude over the horizon in the south. This daily event, the transit of the sun across the celestial meridian, defines **noon**, a fundamental reference in the measurement of time. The interval from one noon to the next sets the length of the **solar day**. Subdivisions of the day have had a long and sordid history, and not everyone welcomed the partitioning of the day into smaller units. As Platus lamented circa 200 B.C.:

> The gods confound the man who first found out how to distinguish hours! Confound him too, who in this place set up a sundial to cut and hack my days so wretchedly into small pieces.

The Romans were the principal partitioners of the day. By the end of the fourth century B.C., they formally divided their day into two parts: before midday (*ante meridiem*, A.M., L. before the meridian) and after midday (*post meridiem*, P.M., L. after the meridian). An assistant to the Roman consul was assigned the task of noticing when the sun crossed the meridian and announcing it in the Forum, since lawyers had to appear in the courts before noon.

By the beginning of the Common Era, the Romans eventually made finer subdivisions of the day. The "hours" of their daily lives were one-twelfth of the time of daylight or of darkness. These variant "hours"—equal subdivisions of the total time of daylight or of darkness—were quite elastic and not really chronometric hours at all. For example, at the time of the winter solstice, by our modern measures there would be only 8 hours, 54 minutes of daylight, leaving 15 hours, 6 minutes for darkness. Near the winter solstice, the Roman daylight "hour" corresponds to

$$\frac{1}{12}(8^h54^m) = 45\frac{1}{2} \text{ minutes}$$

by modern measure.

**Calculate the length (in modern time units) of a Roman "hour" at night at the time of the winter solstice.

At the summer solstice the times were exactly reversed. One-twelfth of a changing interval of time was not a constant from day to day. These "hours" came to be called "temporary hours" or "temporal hours," for they had meaning and length that was only temporary and did not equal an hour the next day. From the Romans' point of view, both day and night always had precisely 12 hours year round. What a problem for the clockmaker!

Sundials were common and were a universal measure of time. They were handy measuring devices since a simple sundial could be made anywhere by anybody without much in the way of special knowledge or equipment. But the cheery boast "I count only the sunny hours" inscribed on many modern sundials, also announces the obvious limitation of the sundial for measuring time. A sundial measures the position of the sun's shadow; thus, no sun, no shadow. And what do you use at night? By the Roman era, water clocks became prevalent and served as a way to measure the shadowless and dark hours. Such clocks had a limited precision, at least by modern standards, but we must be amazed *not* that the Romans did not provide a more precise timepiece but that under their reckoning of hours they were able to provide an instrument that served daily needs at all. It required a hefty amount of ingenuity, but the Romans made their water clocks indicate the shifting length of hours from month to month, rather than from day to day. (The day-to-day changes were too small to be of practical interest.)

The equal hour did not arise until about the fourteenth century. Around 1330 the hour became our modern hour, one of twenty-four equal parts of a day. This new "day" included the night, and it was measured by the *average* time between one noon and the next, the average being over one year. For the first time in history, an "hour" took on a precise, year-round meaning everywhere.
This movement from the seasonal or "temporary" hour to the equal hour is a subtle but profound revolution in human experience. Here was humanity's declaration of independence from the sun, new proof of our growing mastery over ourselves and our surroundings. Only later would it be revealed that we had accomplished this mastery by putting ourselves under the dominion of a machine, with impervious demands all its own.

Procedure

Part A: The Solar and Sidereal Day

Why does the sun appear to move in our sky at all? The earth is spinning about an axis once a day *and* revolving around the sun once a year.

The rotation of the earth about its spin axis once a day has the effect that celestial objects seem to spin around the earth once a day—apart from any motion the celestial objects might have relative to the earth. If the earth did not revolve around the sun but simply spun on its axis, the sun would appear to go around the earth once a day—once a sidereal day. But the earth does revolve around the sun, and this relative motion introduces a small complication: the time from one noon to the next (one solar day) is not the same as the time for the earth to spin once upon its axis (one sidereal day). This difference is so important that we will examine it in two different ways so that its origin is clear.

First, we will adopt a geocentric perspective, that is, we will consider the earth to be motionless and the celestial sphere and the objects on it to revolve around us. For concreteness, let us pick a fixed point on the celestial sphere, say the vernal equinox (see figure 27.1), the intersection of the celestial equator and the ecliptic where the sun goes from south of the equator to north of it. (The sun in its motion along the ecliptic is at the vernal equinox on or about March 21.) Let us further suppose that it is about noon on March 21 so that the sun as well as the vernal equinox are on the local celestial meridian as in figure 27.1a. The celestial sphere rotates (due to, of course, the earth rotating about its spin axis). Sometime during the next day that fixed point on the celestial sphere—the intersection of the ecliptic and celestial equator, the vernal equinox—is again on the meridian. That interval of time is one *sidereal* day. A fixed point on the celestial sphere has gone around once. But in that time the sun has moved eastward along the ecliptic as shown in figure 27.1b. In figure 27.1b, it is not quite noon since the sun is not yet on the meridian. We have to wait a bit longer for the celestial sphere to continue to spin to bring the sun up to the meridian.

1. The sun moves 360° around the ecliptic in 365 days, so the sun moves about 1° per day along the ecliptic. Therefore, the sun is about 1° east of the vernal equinox. Calculate how long you have to wait after the situation depicted in figure 27.1b so that the sun is on the meridian. Express your answer in minutes.

We have described one way of showing that a solar day is a bit longer than a sidereal day. Now we shall examine the situation from a different perspective, a heliocentric one.

In figure 27.2 the earth is initially at position A in its orbit around the sun. It is noon for an observer at the foot of the dotted arrow. Consequently, it is midnight for an observer at the foot of the solid arrow at A. One sidereal day later, the earth is at B and the arrows have the same orientation as at A, but at B the daylight dotted arrow does not point to the sun and so it is not noon. A little later, the earth has turned more upon its axis and moved a bit in its orbit, and now the dotted daylight arrow points toward the sun. The time from A to B constituted one sidereal day while the time from A to C constituted one solar day.

2. Figure 27.3 depicts the sun's apparent motion along the ecliptic (where the zodiacal constellations are found) in a heliocentric perspective. As the earth orbits the sun, the line of sight of the sun and the background stars also move. For example, the diagram shows the sun in Leo. A month later the earth will have moved enough so that Virgo lies behind the sun.

Figure 27.1 (a) A day in late March with the sun and vernal equinox on the local celestial meridian. (b) One sidereal day later Earth has rotated once on its axis and the vernal equinox is back on the celestial meridian. However, the sun has moved eastward along the ecliptic.

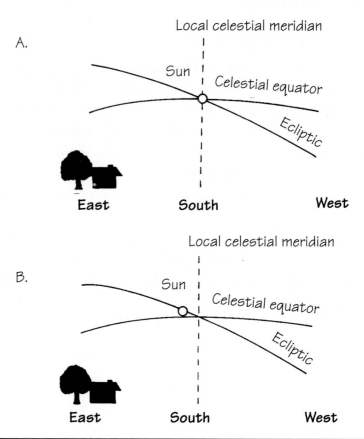

A.

Local celestial meridian

Sun

Celestial equator

Ecliptic

East   South   West

B.

Local celestial meridian

Sun

Celestial equator

Ecliptic

East   South   West

Figure 27.2 The difference between a sidereal day and a solar day arises from the rotation of the Earth about its axis and its revolution around the sun. The illustration is a view of the Earth and sun from far above the north pole of Earth.

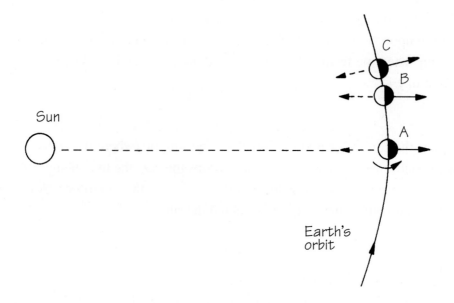

Sun

C

B

A

Earth's orbit

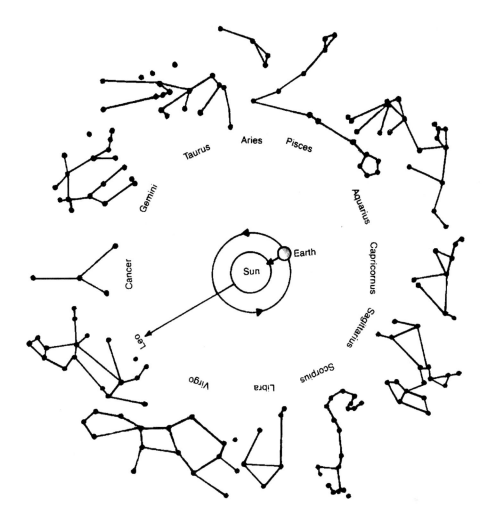

Figure 27.3

Part B: The Analemma—Details of the Sun's Motion

If the earth's orbit were circular and if the earth's spin axis were perpendicular to the plane of its orbit, the sun would always rise precisely in the east, move along the celestial equator at a constant rate, and set precisely in the west. The sun would also appear to move eastward among the stars at a constant rate, completing one revolution in one year. In this idealized situation, the sun would be a perfect clock and would arrive on the observer's meridian at exactly equal intervals.

As you know, the earth's orbit around the sun is elliptical, and the earth's spin axis is tilted 23 1/2° from the perpendicular. These circumstances, as you will see, will cause the time interval between successive meridian crossings of the sun to *vary* throughout the year. We can still refer to a fictitious *mean sun* that moves uniformly along the celestial equator and is on the meridian at noon and again precisely 24 hours later. The real sun, unfortunately, does not behave this way, but the corrections to the time kept by the real sun are not large. They can be represented on a three-coordinate plot called the **analemma** (L., sundial).

The analemma is a closed curve resembling a flatbottomed figure 8 (see figure 27.4). You may have seen the analemma on a terrestrial globe where it is usually placed in the empty part of the Pacific Ocean. Each point on the analemma presents a date in the year. The north-south coordinate at the

203

Figure 27.4 The analemma graphs the sun's declination
and the daily difference between clock noon and noon
by the sundial (sun on meridian for every day of the
year). Declination is the distance in degrees north or
south of the celestial equator.

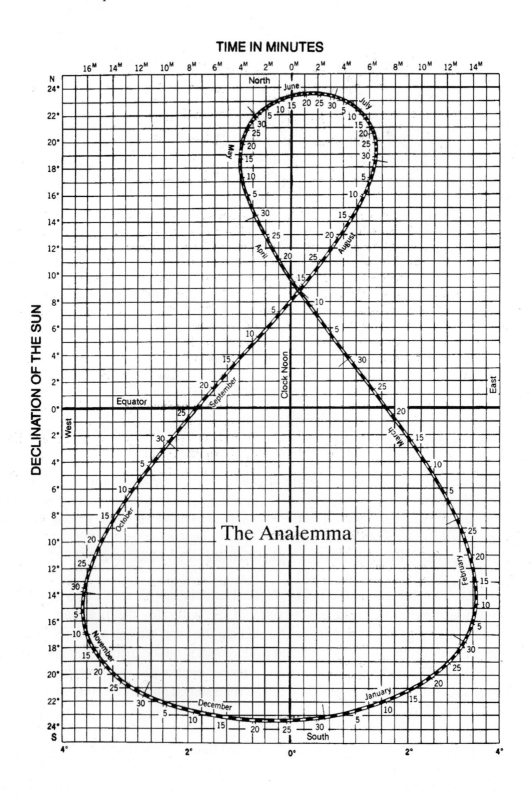

Figure 27.5 Illustration of how the equator-ecliptic angle affects the sun's timekeeping. At the equinox, E represents the solar motion along the ecliptic; its eastward component E' on the equator is shorter. At the solstice, S (equal to E) runs due eastward, and the hour circles are closer together; the component S' is longer than both E' and E.

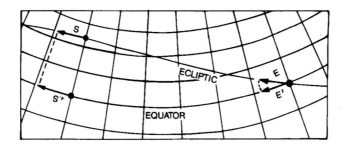

point gives the sun's declination on that date. The east-west coordinate indicates the number of degrees (or minutes of time) by which the sun is east or west of the observer's meridian when the local mean solar time is noon. In the following, we shall examine the origin of the difference between the mean sun and the actual sun and the information contained in the analemma.

The spin axis of the earth is inclined 23 1/2° to the plane of its orbit around the sun. Because of this tilt, the yearly path of the sun eastward among the stars (the ecliptic) is tilted 23 1/2° with respect to the celestial equator. In late June the sun is 23 1/2° north of the equator and in late December 23 1/2° south of it. This annual north-south oscillation of the sun's declination is responsible for the lengthwise extension of the analemma pattern.

The ecliptic tilt has yet another effect on the sun's motion. Since the sun moves along the ecliptic, which is tilted with respect to the celestial equator, the sun's motion relative to the stars is due east only in late June and late December. Hence, the sun's eastward advance per day is greatest at those times and least in March and September when the ecliptic crosses the equator at a slant (see figure 27.5).

Because the meridians of right ascension are more closely spaced at declinations of ±23 1/2° than at the equator, the actual sun's effective eastward motion is faster than that of the mean sun's. Twice a year, near the solstices, the sun arrives later and later on the local meridian because of its relatively fast eastward motion from day to day (look again at figure 27.1 a and b), and as a clock it runs slow. Twice a year near the equinoxes, the sun arrives on the observer's meridian earlier and earlier each day, and as a clock it runs fast. Therefore, two times during the year the actual sun is ahead of clock time and two times during the year it is behind. This effect gives rise to the east-west spread to analemma and determines its general figure-8 shape.

One further influence on the shape of the analemma arises because of the *elliptical* orbit of the earth around the sun. As Johannes Kepler discovered nearly four centuries ago, a planet moves fastest in its orbit near perihelion (point nearest the sun) and slowest at aphelion (point farthest from the sun). Since the earth reaches perihelion on January 3 and aphelion on July 7, the motion of the sun along the ecliptic is faster than average during the winter months and slower than average during the summer months. On January 3, the apparent rate of the sun along the ecliptic is 1.019 degrees per

205

day, while on July 7 the sun moves at a rate of 0.953 degrees per day. The principal effect of this annual velocity variation of the sun is to broaden the southern loop of the analemma and compress the northern loop.

In summary, the analemma graphs the sun's declination, and daily difference between clock noon and noon by the sun (sun on meridian) for every day of the year. Looking at figure 27.4 we see that the sun is west of the mean sun, that is, ahead of clock noon, from September 1 to December 26, falls behind from December 26 to April 15, then moves ahead again until June 15. It falls behind again until September 1, alternately speeding up and slowing down with respect to clock time.

1. Briefly describe the effect on the shape of the analemma if the ecliptic-equator angle were to increase.

2. Obtain from your instructor the latitude of your location and fill it in below.

latitude of your location = _____

Determine the altitude over the southern horizon of the intersection of the celestial equator and the local celestial meridian.

3. Use the analemma in figure 27.4 and your answer to step 2 to fill in table 27.1. The latitude of Altengaard, Norway is +70°.

Table 27.1 Declination and Noon Altitude of Sun			
Date	Declination of Sun	Noon Altitude of Sun at Your Location	Noon Altitude of Sun at Altengaard
February 15			
June 21			
October 15			
December 22			

4. Rip van Winkle awakens from his extended slumber and asks a passerby what year it is so he can determine how long he slept. The passerby quickly responds and then rapidly moves away. Realizing a twenty year snooze might be a world's record, he thinks he should pin down the date as well as the year. He moves out from under the tree where he slept and measures the altitude of the sun when it crosses the meridian; he finds it is 44° over the southern horizon. He knows that the latitude of his chosen spot on the Hudson River in New York is 42°. On what two possible dates could Rip van Winkle have awakened? He notices that buds are appearing on the tree he slept under. What is the date of his awakening?

As we saw previously, the analemma graphs the daily difference between clock noon and apparent noon when the sun crosses the meridian. As an example, check the analemma in figure 27.4 to see that on October 15 the sun will cross the meridian 14 minutes before clock noon.

5. For the dates given in the table below, use the analemma in figure 27.3 to determine whether the actual sun will cross the meridian before or after clock noon and by how many minutes.

Table 27.2 Sun Time and Clock Time		
Date	Sun Crosses Meridian Before or After Clock Noon	Amount of Time Before or After Clock Noon
January 10		
February 15		
April 15		
July 30		
September 1		
October 30		

6. Determine the altitude over the southern horizon of the actual sun and the time when it crosses your local meridian on the dates in the table below.

Table 27.3 Sun Altitude and Time of Meridian Crossing		
Date	Altitude of Sun	Time When Sun Crosses Meridian
May 15		
June 15		
November 4		
December 25		

Experiment 28: Diffusion and Osmosis

Invitation to Inquiry

Diffusion occurs when there is a concentration gradient and molecules are in constant motion. This means that they will move from a place where they are in a higher concentration to a place where they are in lower concentration.

Try the following:

1. Pick a quiet room for this work. It could be in a house, a lab, or a small room such as your dorm room.
2. Be sure that no one will be moving about or coming in during your investigation.
3. Pick a place in the room where you can locate a container of one of the following: an aromatic liquid such as household cleaner (e.g., Formula 409), perfume/aftershave, or a solid such as a spice or a scented candle.
4. Be sure to have a stopwatch or one which can measure seconds.
5. On the first day place the aromatic material in your selected spot.
6. Open the top of the container and start timing your work.
7. Quickly move to a pre-selected spot in the room as far from the open container as possible.
8. Stay in that place until you sense the aroma of the material.
9. Note this time.
10. Repeat this procedure with other materials on different days under similar conditions.
11. Record the time it takes for you to smell these materials. These are the diffusion rates of your various aromatic materials.

Once you have gathered all your data, compare your results. What factors could contribute to differences in the diffusion rates? Consider such factors as (1) the phase of the material (solid, liquid, gas), (2) molecular weight, (3) solubility, and (4) air currents. Why should you do this in several different days?

Background

Although you may not know what diffusion is, you have experienced the process. Can you remember walking into the front door of your home and smelling a pleasant aroma coming from the kitchen? It was diffusion of molecules from the kitchen to the front door of the house that allowed you to detect the odors. **Diffusion** is defined as the net movement of molecules from an area of greater concentration to an area of lesser concentration until the concentration everywhere is the same. The movement in one direction minus the movement in the opposite direction determines the

direction of net movement. To better understand how diffusion works, let's consider some information about molecular activity.

The molecules in a gas, a liquid, or a solid are in constant motion because of their kinetic energy. Moving molecules are constantly colliding with each other. These collisions cause the molecules to move randomly. The higher the concentration of molecules in one region, the greater the number of collisions. Some molecules are propelled into the less concentrated area, and others are propelled into the more concentrated area. Over time, however, there will be more collisions in the highly concentrated area, resulting in more molecules being propelled into the less concentrated area. Thus, the net movement of molecules is always from more tightly packed areas to less tightly packed areas.

Diffusion occurs when there is a difference in concentration from one region to another or from one side of a membrane to another (figure 28.1a). A difference in the concentration of molecules over a distance is called a **concentration gradient**. When the molecules become uniformly distributed, as in figure 28.1b, they have reached **dynamic equilibrium**, in which the number of molecules moving in one direction is balanced by the number moving in the opposite direction. It is dynamic because molecules continue to move, but because motion is equal in all directions and there is no net change in concentration over time, equilibrium exists. The process of diffusion occurs in both living and nonliving systems. Biologically speaking, diffusion is responsible for the movement of a large number of substances, such as gases and small uncharged molecules, into and out of living cells.

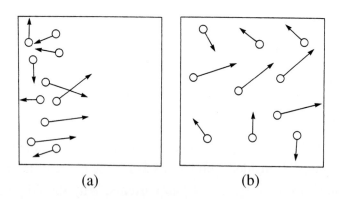

(a)   (b)

Figure 28.1

The direction of diffusion is always from where there were originally more molecules to where there are fewer. This is similar to the scattering of a crowd of people leaving a theater. Many of the individuals move from the theater to the outside, but some go back to retrieve their gloves or popcorn. The net movement, however, is the movement of the individuals leaving the theater minus the movement of those returning.

Imagine that your instructor opens a bottle of ammonia in a corner of the room. The bottle would have the highest concentration of ammonia molecules in the room; the individual ammonia molecules would move from this area of highest concentration to where they are less concentrated (figure 28.2).

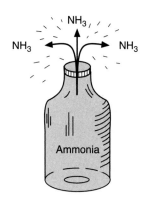

Figure 28.2

Although you could not actually see this happening, ammonia molecules would leave the bottle and move throughout the air in the room because of molecular movement. You could detect this by the odor of the ammonia. If you compare the relative number of ammonia molecules in the bottle to those dispersed in the room, you would be dealing with what is called relative concentration. **Relative concentration** compares the amount of a substance in two separate locations. Whenever there is a difference in concentrations of a substance, you can predict the direction that most of the molecules will move. You can predict that when the bottle is first opened, ammonia molecules will move from the area of higher concentration (the bottle) to the region of lower concentration (the air in the room). Soon, however, the molecules of ammonia will mix with the air molecules in the room. Because the ammonia molecules are moving randomly, some of them will move from the air back into the bottle.

As long as there is a higher concentration of ammonia molecules in the bottle, more of them move out of the bottle than move in. One way of dealing with the direction of movement is to compare the number of molecules leaving the bottle with the number reentering the bottle. This is called the net amount of movement. The movement in one direction minus the movement in the opposite direction is the direction of net movement. If, for example, 100 molecules of ammonia leave the bottle and 10 reenter during that time, the net movement is 90 molecules leaving the bottle. Ultimately, the number of ammonia molecules moving out of the bottle will equal the number of ammonia molecules moving into it. When this point is reached, the ammonia molecules are said to have reached dynamic equilibrium.

When several kinds of molecules are present, consider only one case of diffusion at a time even though several different types of molecules are moving. For example, consider the exchange of gases between the lungs and blood. In the lungs, there are a series of tubes that transport gases. These tubes divide into smaller and smaller branches and eventually end at a series of small alveolar sacs. Adjacent to these sacs are a number of capillaries containing blood. By the process of diffusion, there is an exchange of oxygen and carbon dioxide between the alveolar sacs and the blood in the capillaries (figure 28.3).

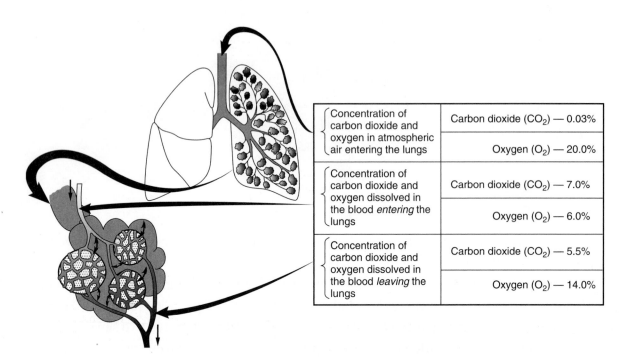

Concentration of carbon dioxide and oxygen in atmospheric air entering the lungs	Carbon dioxide (CO_2) — 0.03%
	Oxygen (O_2) — 20.0%
Concentration of carbon dioxide and oxygen dissolved in the blood *entering* the lungs	Carbon dioxide (CO_2) — 7.0%
	Oxygen (O_2) — 6.0%
Concentration of carbon dioxide and oxygen dissolved in the blood *leaving* the lungs	Carbon dioxide (CO_2) — 5.5%
	Oxygen (O_2) — 14.0%

Figure 28.3

1. Is the direction of net movement of carbon dioxide molecules in figure 28.3 from the blood to the lungs or from the lungs to the blood? Explain your answer.

2. Is the direction of net movement of oxygen molecules in figure 28.3 from the blood to the lungs or from the lungs to the blood? Explain your answer.

Another example of diffusion is sugar dissolving in water. When sugar molecules and water molecules mix, a solution is created. A solution is any mixture where two or more different types of molecules are evenly dispersed throughout the system.

3. Draw an arrow on figure 28.4 to show the net direction of sugar movement.

Figure 28.4

Figure 28.5 shows a differentially permeable membrane. A **differentially permeable membrane** is a thin sheet of material that selectively allows certain molecules to cross but prevents others from crossing. The membrane in this figure is permeable only to water molecules. Water molecules may freely diffuse across the membrane, but other types of molecules cannot. The diffusion of water across a differentially permeable membrane is called **osmosis**. On each side of the differentially permeable membrane in figure 28.5 is a chloride solution. A solution is characterized by the dissolved substance called the solute. Chloride in this example is the solute. The substance in which the solute is dissolved is called the solvent. In figure 28.5 and in biological systems, water is the solvent.

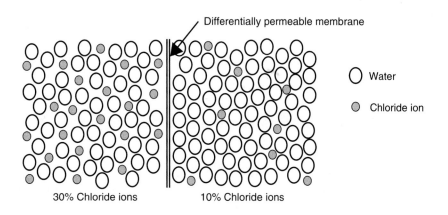

Figure 28.5

4. What is the percentage of solute in the left side of the container (figure 28.5)?

5. What is the percentage of solvent in the left side of the container?

6. What is the percentage of solute in the right side of the container?

7. What is the percentage of solvent in the right side of the container?

8. Where is the water in higher concentration—the left or right side?

9. Draw an arrow to indicate the net direction of movement of the water molecules.

In each of the previous examples, the net movement was a result of diffusion of molecules from a place of higher concentration to a place of lower concentration. The rate at which diffusion occurs is related to the amount of energy the molecules have and the degree of difference between the areas of high and low concentration. Adding energy doesn't change relative concentrations, nor does it influence the direction of diffusion. It merely affects the rate at which diffusion occurs. Molecules with greater kinetic energy move faster causing diffusion to happen quicker.

The kinetic molecular theory states that all substances are made up of molecules that occupy space and are constantly in motion. This exercise helps you examine some phenomena related to this motion of molecules.

During this lab exercise you will

1. set up a demonstration of osmosis under a variety of temperature conditions and determine how temperature influences the rate of osmosis.
2. set up a demonstration of osmosis using a variety of concentration gradients and determine how concentration differences influence the rate of osmosis.
3. graph the results of the osmosis demonstrations.

Procedure

Part A: Osmosis and the Effect of Temperature

1. Working in groups, prepare three sacs to demonstrate osmosis.
2. Obtain three pieces of dialysis tubing (sausage casing) and soak them in tap water for about 1 minute.
3. Form each of the pieces of tubing into a tubular bag. Shake off the excess water, fold over one end of the dialysis tubing, and securely tie it with a piece of string.
4. Fill each tubular bag about half full with full-strength molasses. Leave room for a small pocket of air. Tie the open end of each bag. After rinsing each bag, pat it dry and cut off any excess string.
5. Use a balance to obtain an initial weight for each bag. Record this data in Data Table 28.1.
6.   a. Place one bag in a beaker of water at room temperature (approximately 20°C).
       b. Place the second bag in a water bath heated to 40°C.
       c. Place the third bag in a beaker of ice water (approximately 0°C) (figure 28.6a). Record the

214

exact temperature of the water in each beaker. Make certain that each bag is completely covered with water.

7. After 5 minutes, remove each bag and gently squeeze to assess any changes in firmness. Observe the size, shape, and firmness of each bag. Gently pat each bag dry and weigh. Record your results in Data Table 28.1.

8. Return each bag to its appropriate container of water for another 5-minute interval. Repeat your observations and measurements every 5 minutes. (Measurements should be taken at 0, 5, 10, 15, and 20 minutes.)

9. To more easily visualize the effect of temperature on osmosis use the data collected in Data Table 28.1 to construct a graph of the effect of temperature on the rate of osmosis. Graph paper is provided on page 219. If you do not have three different colored pencils, use a solid line for the hot water, a dashed line for room-temperature water, and a dotted line for the ice water.

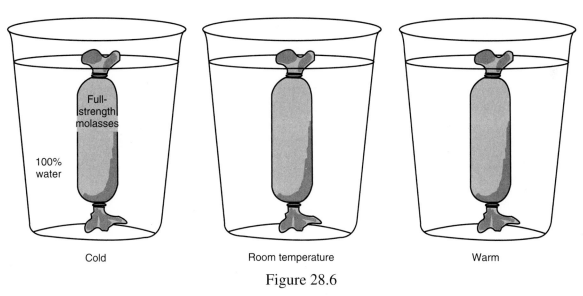

Figure 28.6

Data Table 28.1	Effect of Temperature on the Rate of Osmosis				
Temperature	Initial weight	5 minutes	10 minutes	15 minutes	20 minutes
Warm					
Tap					
Ice					

215

Part B: Osmosis and the Effect of Concentration

1. Repeat the construction of the differentially permeable bags (see instructions under "Osmosis: Effect of Temperature"). Prepare three bags but fill each with a different concentration of molasses according to the following specifications (figure 28.7b):

> Bag 1: one part molasses to three parts water (2.5 ml molasses; 7.5 ml water).
> Bag 2: one part molasses to one part water (5 ml molasses; 5 ml water).
> Bag 3: full-strength molasses (10 ml molasses).

2. Rinse and gently pat the bags dry.
3. Record the initial weight of each bag.
4. Place each bag in a container of room-temperature tap water.
5. Check their weight and firmness at 5 minute intervals for a period of 20 minutes.
6. Record all data in Data Table 28.2 and graph your results on the graph provided on page 221.

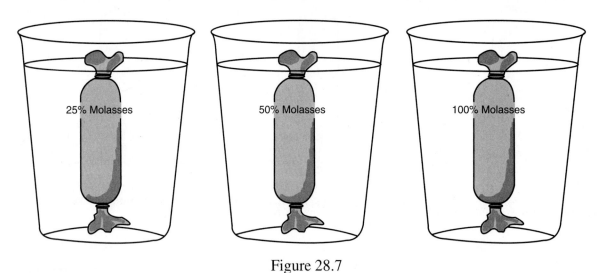

Figure 28.7

Data Table 28.2	Effect of Concentration on the Rate of Osmosis				
Concentration	Initial weight	5 minutes	10 minutes	15 minutes	20 minutes
25%					
50%					
100%					

Results

1. In a perfectly tied and unbroken bag, should we see evidence of sugar molecules passing through the "membrane"? Qualify your answer in terms of how differential permeability operates.

2. From your graph of the influence of temperature on the rate of osmosis, what can you conclude about the effects of temperature on the rate of osmosis?

3. Was dynamic equilibrium reached in any of the molasses demonstrations? Explain your answer.

4. How do differences in concentration affect the rate of osmosis?

5. Why does a good cook wait to put the salad dressing on a salad until just before serving? Answer by explaining what happens to the cells in a lettuce leaf when the dressing is added.

6. Human cells contain 0.9% solute (dissolved materials). Therefore, there is 99.1% water in these cells. The Pacific Ocean contains 3.56% salt. Although this seems like a silly question, how much (%) water is in the ocean? You are cast adrift on this ocean. What would happen to your cells if you were to drink the salt water?

7. Your younger brother just put your favorite saltwater fish into his freshwater aquarium. Predict what will happen to the fish, its cells, and your younger brother.

8. Was the purpose of this lab accomplished? Why or why not? (Your answer to this question should show thoughtful analysis and careful, thorough thinking.)

Osmosis: Effect of Temperature

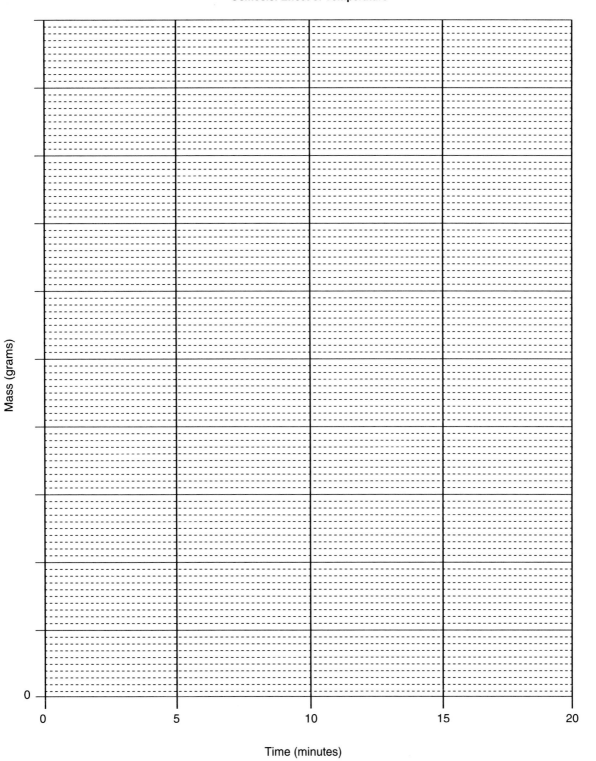

Mass (grams)

0

Time (minutes)

0   5   10   15   20

Osmosis: Effect of Concentration

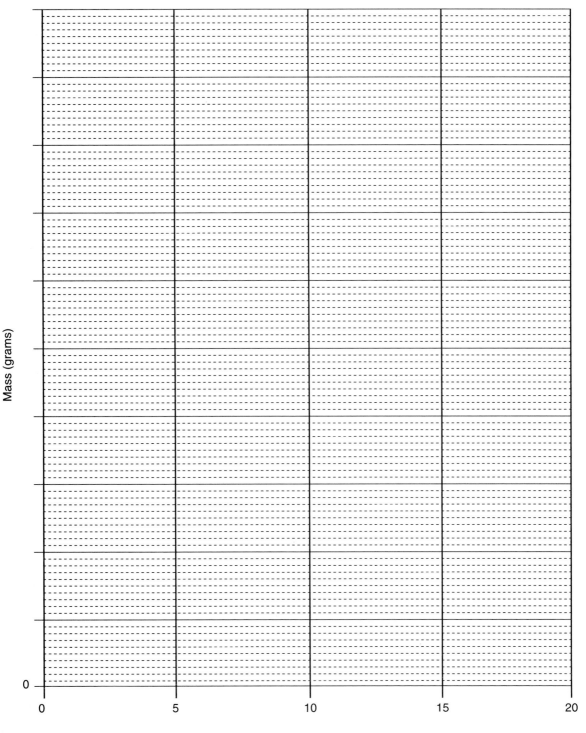

Mass (grams)

0

Time (minutes)

0   5   10   15   20

Experiment 29: The Microscope

Invitation to Inquiry

The jobs of a microscope are to magnify and resolve; for example, it should enlarge and should show you the details of the component parts of the material you choose to observe. This lab exercise will provide you with the opportunity to use the most common microscope, the compound light microscope. There are many other kinds of microscopes that play vital roles in providing information about the microscopic world about you.

Go to the internet and dig out information that will allow you to compare and contrast different kinds of microscopes including the polarizing, electron, and tunneling microscopes. Compare how each functions in relation to the compound light microscope.

Background

Because biological study includes the microscopic examination of one-celled organisms and the cells and tissues of multicellular organisms, it is important to learn the correct procedures for efficient operation of a light microscope. In addition to giving you an opportunity to learn proper microscope technique, this exercise also gives you a chance to practice using a microscope.

During this lab exercise you will
1. identify and name the parts of a light microscope and describe the functions of the various parts.
2. determine the total magnification of a set of lenses in a compound microscope.
3. focus on a practice slide.
4. focus on crossed hairs in a temporary wet mount to determine the depth of field of a set of lenses.

Procedure

Take your assigned microscope to your work station. Refer to figure 29.1 and table 29.1 to familiarize yourself with the operation and function of each microscope part.

Magnification

The **compound microscope** is a device that uses two sets of lenses to increase the apparent size of objects. The two lenses are known as the **ocular lens** (the lens you look through) and the **objective lens**, which is near the stage. The **magnifying power** (how much it magnifies) of each lens

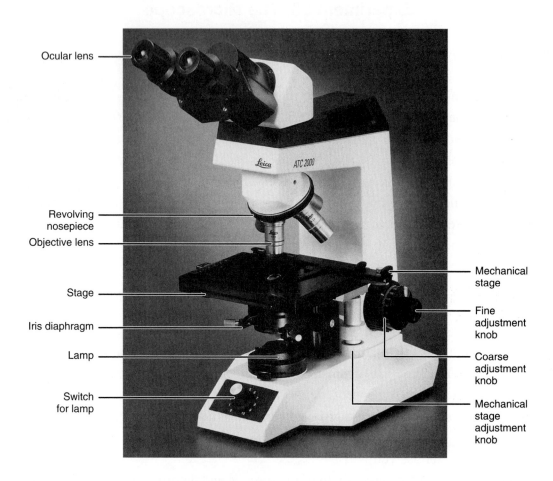

Ocular lens

Revolving nosepiece

Objective lens

Stage

Iris diaphragm

Lamp

Switch for lamp

Mechanical stage

Fine adjustment knob

Coarse adjustment knob

Mechanical stage adjustment knob

Figure 29.1 (Courtesy of Leica, Inc.)

Table 29.1   Parts and Function of a Light Microscope

Part	Function
Ocular lens (eyepiece with pointer)	Lens through which you view magnified specimen. Pointer may appear as a needle or as a curved line.
Revolving nosepiece	Movable mount for selecting the objective lens which provides the magnification desired.
Objective lens	Lens on revolving nosepiece, which accomplishes the initial magnification of the specimen.
Stage	Flat work surface upon which the slide is placed.
Iris diaphragm	Regulates the amount of light passing through the stage aperture and specimen.
Lamp	Constant light source beneath the iris diaphragm.
Mechanical stage	A microscope with a mechanical stage has a lever that is opened laterally (never lifted) to accept and secure the slides to the stage. Control knobs are used for precise movement of a slide on the stage.
Coarse adjustment knob	Gives initial focus on low power.
Fine adjustment knob	Gives refined focus on low power, high power, and oil immersion.

is marked on its tubular housing. Simply multiply the magnifying power marked on the ocular lens housing times the value marked on the objective lens housing to determine how many times your specimen is enlarged. Notice that your ocular lens magnification is 10. If the low-power lens is also marked 10, the total low-power magnification is $10 \times 10 = 100$.

Microscopes often have additional objective lenses, namely a **scanning lens**, which typically has a magnifying power of 4 and is used for initial viewing of the specimen; a **high-power lens**, which typically has a magnifying power of 45; and an **oil-immersion lens**, which typically has a magnifying power of 100. As the magnifying power increases, the lenses get longer. *Use of the oil-immersion lens requires special training, so do not use it unless instructed to do so by your instructor.* Improper use could cause severe and costly damage to the oil-immersion lens.

Calculate and record here the magnification of your microscope when the high-power objective lens is used.

Resolving Power

Resolving power is a measure of lens quality. Quality lenses have a **high resolving power**, which is the capacity to deliver a clear image in fine detail. If a lens has a high magnifying power but a low resolving power, it is of little value. Although the image may be large, it is not clear enough to show fine detail.

Another factor that influences resolving power is the cleanliness of the lenses. Dirt, water, or oil on the lens may scatter light and reduce the effective resolving power of the microscope. Therefore, lenses should always be kept clean. *Use only lens paper to clean the lenses.*

Field of View

You have already learned that lenses can have different magnifying powers, but it is also important to understand that each lens has a particular field of view. The **field of view** is the size of the area that the lens views. *The larger the magnifying power of an objective lens, the smaller the area viewed.* This is sometimes hard to appreciate because to you—the observer—the size of the circle of light you see through the ocular lens appears the same for all powers of the objective lenses. When you switch from low power to high power, you are actually looking at the central portion of what was visible under low power. Therefore, it is important to *center the specimen on low power before making the switch to high power.*

Parfocal Capability

A feature of a good quality microscope is its **parfocal capability**. This means that when a specimen is in focus under low-power magnification, you can switch to high-power magnification and have the specimen remain in reasonably good focus. Usually, just a slight touch to the fine adjustment knob is all that is needed to sharpen the focus. Of course, it is imperative that your specimen be accurately centered just before you switch over to high power.

Viewing and Focusing

Before you attempt to view any specimen through the microscope, you must learn the correct use of its parts.

Always use the following procedures when viewing objects through the microscope:

1. Carefully carry your assigned microscope to your work space using both hands. One hand should hold the microscope by the arm and the other hand should support the microscope base.

2. Make sure the microscope is plugged in. Turn on the light.

3. Rotate the low-power objective lens or scanning lens into position directly over the round opening in the flat stage of the microscope. Move it until it clicks into place.

4. Rotate the coarse adjustment knob so that the distance between the objective lens and the stage is at its maximum.

5. Center your specimen over the opening in the stage. Make sure that the slide is securely held in place by the clips or the the fingers of the mechanical stage.

6. Watch the stage and objective lens from alongside (not through) the microscope. Make the distance between the specimen and the low-power lens as small as possible.

7. While looking through the ocular lens, turn the coarse adjustment knob to move the objective lens away from the specimen until a part of the specimen comes into focus.

8. The iris diaphragm is located below the stage and regulates the amount of light passing through the specimen. Locate the lever that adjusts the iris diaphragm and move it so that you get a good image.

9. While looking through the ocular lens, center the specimen in the field of view.

10. Switch to high power and sharpen the focus with the fine adjustment knob only.

11. If you are unable to find the specimen, switch back to low power and repeat steps 7, 8, and 9.

12. Keep both eyes open even though only one is used in the monocular, compound microscope. After a short while, you can get accustomed to ignoring impressions coming from your free eye. If you have trouble at first simply cover your free eye with your hand. Squinting leads to muscle fatigue and headaches.

Making a Wet Mount

A wet-mount slide is made by placing the object in a drop of water on the slide and covering it with a thin glass coverslip. A coverslip must always be used. The use of a coverslip gives a flat surface to look through. If you don't use a coverslip, the water drop will form a curved surface and make viewing difficult. In addition, on high power the heat from the lamp will cause water to evaporate and condense on the lens. A fogged lens is difficult to see through.

1. Make a wet-mount slide by cutting out one of the words in figure 29.3; your instructor may indicate a particular word.
2. Place the word on the slide, put one or two drops of water on the paper, and place a coverslip over the paper. If you place one edge of the coverslip against the glass slide and gently lower it into position, as shown in figure 29.2, you will not trap air bubbles which interfere with your ability to see the object.

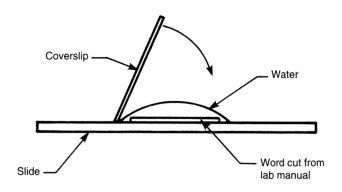

Figure 29.2

3. Place this slide on the microscope and examine it under low power. You should find that something comes into view with about a quarter of a turn of the coarse adjustment knob. Move the knob smoothly and slowly. If you cannot see anything, start over by returning the low-power lens to the position closest to the slide and trying again. If you still have trouble, ask your instructor for assistance. The first things you see are the fibers in the paper. You may also see the ink that forms the letters on the paper. If you do not see any letters, move your slide around until you do.
4. Adjust the amount of light by moving the lever connected to the iris diaphragm. Notice that there is an optimal position for this lever that allows you to see the letters on the paper clearly. If you have any problems locating something to look at, call your instructor.

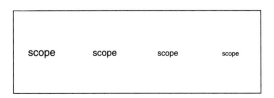

Figure 29.3

227

5. Center a letter or portion of a letter in the center of the low-power field of view and switch to high power.
6. Focus using the fine adjustment knob.
7. Use a pencil to sketch the word as seen with (1) the unaided eye, (2) the low-power objective, and (3) the high-power objective of the microscope. The letters in your drawings should be oriented as they actually appear when viewed through the microscope.

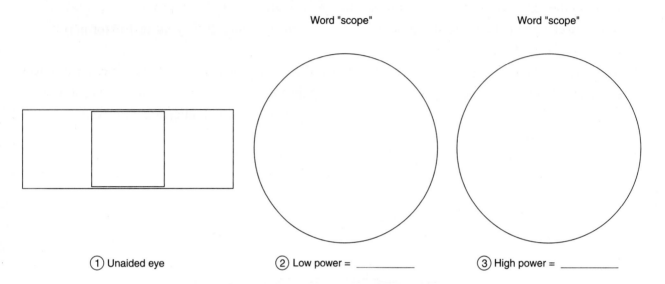

Word "scope"   Word "scope"

① Unaided eye   ② Low power = _____   ③ High power = _____

a. Compare your three drawings. How are they different in regard to the number of visible letters?

b. As you move the slide slightly to the right, in which direction does the letter appear to move?

c. In what direction would you move the slide if a swimming specimen were leaving the field of view at the top of the slide and you wanted to continue looking at it?

Planes of Focus and Depth of Field

Lenses have a **plane of focus** which is a position a specific distance from the lens where an object (or a portion of it) appears in sharp focus. You can actually move up and down through some specimens and focus on different levels of the specimen. The portion that is in focus at any given level is the portion in the plane of focus. This plane of focus has some depth to it. Therefore, it is a layer in space, a specific distance from the lens where objects are in sharp focus. The thickness of the layer that is in sharp focus is known as the **depth of field**. High-power lenses have a much shallower depth of field than do low-power lenses.

To experience microscopic depth of field, prepare a wet mount of two human hairs. If possible, use two different types of hair—light hair and a dark hair or a coarse hair and a fine hair.

1. Place a drop of water on the center of a clean slide.
2. Take a piece of hair (about 2 cm long) and place it lengthwise in the drop of water. Then take a second piece of hair (again, about 2 cm long) and lay it over the first so that they are crossed at right angles.
3. Focus on the crossed hairs under low power. You will probably see both hairs clearly at the same time (figure 29.4). They are both within the depth of field.

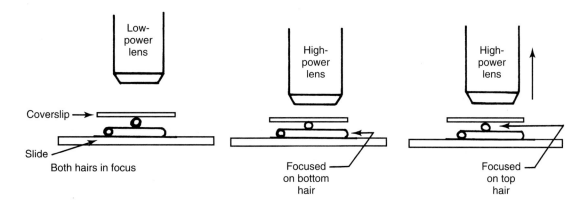

Figure 29.4

4. Move the slide so that the point where the hairs cross is in the center of the field of view and switch to high power. Use the fine adjustment knob to bring the bottom hair into focus. Note that only one hair is in sharp focus. It is within the depth of field. The other is blurry. It is still distinguishable as a hair but is not clearly focused. Focus upward with the fine adjustment knob to see that the bottom hair goes out of focus as the top hair comes into focus.

Is the high-power depth of field greater or less than the low-power depth of field?

Further Practice

1. Obtain a prepared slide. These slides have been stained with particular stains so that specific structures are highlighted. If the slide is of a large object, it has been sliced into thin sheets which were then mounted on the slide, and a coverslip was permanently affixed to the top surface. Remember, your purpose is to practice looking at microscopic structures. Use the iris diaphragm, low power, and high power, and move the slide around. Make drawings of your observations in the following space. Be sure to label all drawings with the name of the organism and the total magnification as demonstrated in the following example. Make your drawings large so that you can show details of the structure.

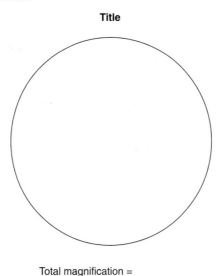

Title

Total magnification =

2. Prepare wet mounts of the fresh specimens available. This variety of material could include such things as protozoa, cork, potato, algae, or microscopic animal life. Draw your observations here.

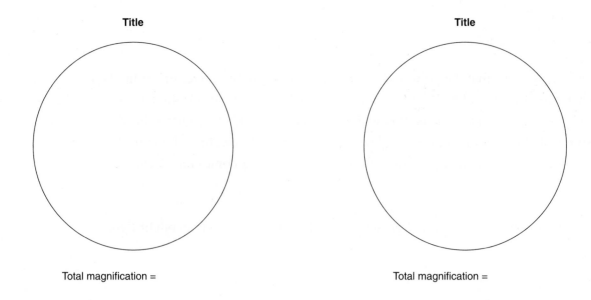

Title

Total magnification =

Title

Total magnification =

3. When finished, clean and dry your slides and return them to the slide box. Since coverslips are fragile, your instructor will inform you how to clean and return them. If you break any slides or coverslips *do not put them in the waste paper basket*. Place broken glass in the container designated for the purpose.

4. When you are finished using your microscope make sure you have removed any slides, clean any moisture from the stage or lenses, and position the nosepiece on low power before returning the instrument to its correct storage place.

Results

1. Part of learning how to use the microscope is learning how to troubleshoot. When you have difficulty, what should you do? In each of the following situations, list possible causes of the problem and indicate what you should do to resolve it.

   a. You have just set your microscope up for use and turned it on. You cannot see a field of view. Possible problem:

   Solution:

   b. Your slide is in focus on your microscope, but you do not see the specimen. Possible problem:

   Solution:

   c. The specimen that you are examining is very thin and transparent. What can you do to make it easier to see?

   d. You are looking at a specimen on low power and you cannot see the details that your instructor asked you to look at. What should you do? How is this done?

2. Complete the table, indicating the magnifying power of your microscope.

	Ocular lens	Objective lens	Total magnification
Scan			
Low			
High			

3. The following diagrams illustrate the field of view if you were using low-power magnification. Circle the part of the slide (numbers) you would see when switching from low- to high-power magnification.

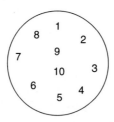

4. If you were using low power and wanted to look at the number 3 on high power, what should you do before you switch to high power?

5. Label the following structures on the microscope drawing on the next page: lamp, fine adjustment knob, coarse adjustment knob, mechanical stage knob, revolving nosepiece, stage, ocular lens, objective lens, iris diaphragm.

6. Was the purpose of this lab accomplished? Why or why not? (Your answer to this question should show thoughtful analysis and careful, thorough thinking.)

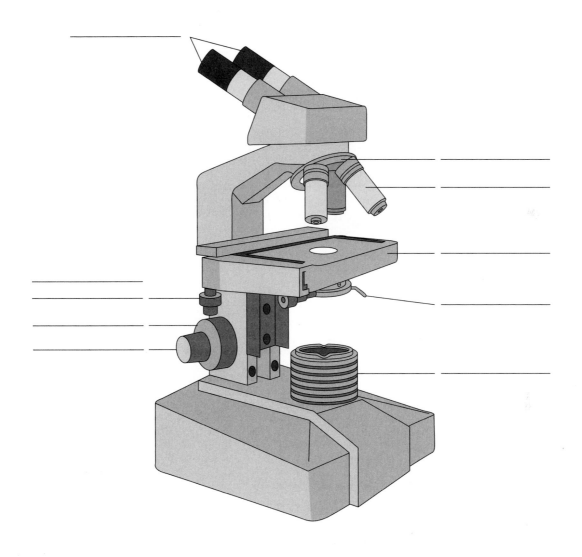

(From Ruth Bernstein, et al., *Biology Laboratory Manual*. Copyright 1996 The McGraw-Hill Companies, Inc. New York, NY. All Rights Reserved. Reprinted by permission.)

Experiment 30: Survey of Cell Types: Structure and Function

Invitation to Inquiry

Understanding the structure of the typical eukaryotic cell can be a real challenge. To help, you can construct a model using materials readily available to you. Based on the material present in the lab and the text, search out "ingredients" such as spaghetti, paper clips, ping-pong balls, and plastic wrap, and arrange them into a model of a typical eukaryotic cell. When finished take them to your instructor and explain your reasons for using the materials you have chosen to be sure that you on the right track. This might seem like a silly thing to do, but you may be very surprised to find just how much it helps in your learning of cell structure.

Background

The cell concept is basic to understanding the activities and characteristics of organisms. **Cells** are the smallest units of living things and are the units of structure and function of an organism. As functional units, they reflect the abilities of the organism as a whole. Some simple kinds of organisms consist of individual cells, but many of the organisms with which we are most familiar are multicellular. Multicellular organisms usually are composed of several different kinds of cells, each having specific characteristics that relate to its function. The various kinds of living things have been subdivided into three domains: Eubacteria, Archaea, and Eucarya. The cells of the Eubacteria and Archaea are small and simple and lack a nucleus. These cells are called **prokaryotic cells**. The cells of the Eucarya all have a **nucleus** and other kinds of structures called **organelles** within the cell. This type of cell is called a **eukaryotic cell**. Because the Archaea and Eubacteria are extremely small and difficult to see, in this exercise we will look only at Eubacteria as examples of a prokaryotic cell. We will spend the majority of our time looking at the various classifications of the Domain Eucarya, which is divided into the following kingdoms: Protista (algae and protozoa), Fungi, Plantae, and Animalia.

Robert Hooke was the first person to use the word cell in reference to the units that make up organisms. He examined cork under a microscope and saw the cell walls of these plant cells. Hooke recounts his important observation:

I took a good clear piece of cork, and with a Penknife sharpen's as keen as a Razor, I cut a piece of it off . . . then examining it very diligently with a Microscope . . . I could exceeding plainly perceive it to be all perforated and porous, much like a Honey-Comb in these particulars . . . in that these pores, or cells, were not very deep, but consisted of a great many little Boxes. . . . For, as to the first, since our Microscope informs us that the substance of Cork is altogether fill'd with Air, and that Air is perfectly enclosed in little Boxes or Cells distinct from one another.[1]

[1]Source: Robert Hooke (1635-1703), quoted in Gabriel and Fogel, *Great Experiments in Biology*. Copyright 1955 by Prentice Hall, Inc.

Today we recognize that Hooke saw only the cell walls of plant cells. However, he did recognize that living material was made of many similar subunits which he called cells. We continue to use his terminology today.

During this lab exercise you will

1. prepare a temporary wet mount of sections of onion membrane, view the specimen through a microscope, identify common structures, and make a three-dimensional drawing of a typical onion cell.
2. make a temporary wet mount of an *Elodea* leaf and view its cellular structure through a microscope, identify common structures, and make a three-dimensional drawing of a typical cell.
3. make a temporary wet mount of *Spirogyra*, *Euglena*, and *Paramecium*, view the cells through a microscope, identify common structures, and make a three-dimensional drawing of a typical cell.
4. observe cheek epithelial cells through a microscope, identify common structures, and make a three-dimensional drawing of a typical cell.
5. observe slides of fungi, soil bacteria, and *Anabaena* and note their structures and characteristics.

Procedure

Kingdom Plantae

We will begin with plant cells because they are relatively large and have several organelles that can be easily identified. Plants have many different kinds of cells organized into complex structures like leaves, fruits, and stems. We will look at two examples of plant cells: onion and *Elodea*.

Onion

An onion is composed of overlapping layers which form rings when the onion is sliced. Cut a small piece of an onion ring approximately 1 cm × 1 cm. On the concave surface of the piece of onion is a thin membrane that consists of many onion cells attached to one another. This membrane is one layer of cells thick. Peel this membrane from the rest of the piece of onion. Be careful to not wrinkle it and place it in a drop of water on a slide. Place a coverslip over the entire preparation and examine under a microscope. It should look something like figure 30.1. Begin viewing the onion cells with low power and proceed to high power to see detail.

You should be able to see the following structures that are typical of plant cells: (1) cell wall, (2) nucleus, (3) one or more nucleoli in the nucleus, (4) a large central vacuole, and (5) cytoplasm. The **cell wall** is found on the outside of the cell and provides a "box" within which the rest of the cell is found. The cell walls of plant cells are composed of a complex carbohydrate known as cellulose.

The **nucleus** will appear as a round or egg-shaped structure inside the cell, and the small structures seen inside the nucleus are the **nucleoli** (singular **nucleolus**). The vacuole and cytoplasm will probably be the most difficult to recognize. The **cytoplasm** will appear as a granular material near the cell wall. This will be a very thin layer. There is an outer boundary to the cytoplasm known as the **cell membrane** which is located inside the cell wall and outside the cytoplasm. However, it is very thin, and in plant cells it is pressed up against the cell wall making it difficult to see. Often in unstained cells, which are still alive, the cytoplasm may be seen to flow. Look closely at the granules or specks in the cytoplasm to see if they are moving. These granules are objects or cell structures too small to be seen clearly with the light microscope. The **vacuole** is a large water-filled space in the center of the cell. Because it does not have any particles in it, it appears to be empty, but it is not. These cells are not flat but resemble a structure similar to a shoebox. The box itself represents the cell wall, the space in the center represents the vacuole, and all the other structures (nucleus and cytoplasm) are squeezed in between the cell wall and the vacuole.

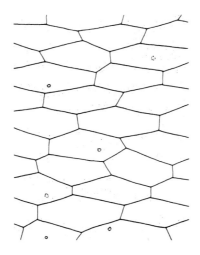

Figure 30.1

Staining

After you have examined the wet mount of the living onion cells, you can stain the cells to make some of the structures easier to see. Biologists often use **stains** that bind to various macromolecules and structures in the cell to make the structures more visible. Living cells may be stained to show cilia, flagella, the nucleus, or other cell organelles. Some stains destroy cells immediately, whereas others, called "vital stains," kill cells more slowly. The organisms absorb these stains and continue to carry on their life functions for some time.

1. *Use caution when using all stains. Many will stain your hands and clothing.* Use Lugol's solution (composed of iodine and water) to stain your onion tissue, as demonstrated in figure 30.2. Lugol's solution stains carbohydrates such as starches and glycogen.

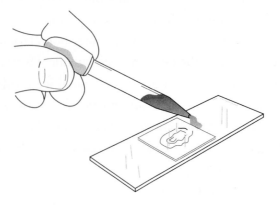

(a) Add one or two drops of stain to edge of coverslip.

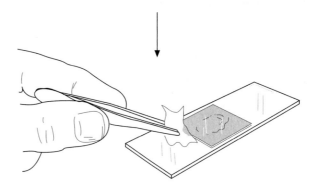

(b) Draw the stain under by touching lens paper to the opposite side of coverslip.

Figure 30.2

Microscopic Drawings

The average student is not gifted with a "photographic mind." Many observations, including those done using a microscope, must therefore be recorded for later study and review. That means drawings must be made! Make drawings in pencil so that you can make modifications easily. Biological drawings should be simple but accurate representations of your observations. Make your drawings large enough so that you can show clear details. You are not expected to be an artist; such drawings are for your benefit. Drawings should be labeled with a title and the total magnification used. As you make sketches you will look more closely at the cells, and this will help you remember what you saw. If you have seen an object well enough to reproduce it accurately in a drawing, then you have seen it well.

1. Sketch the three-dimensional shape of an onion cell in the outline in the space provided. Draw the cell structures in their proper relationships to one another as you viewed them and label the following: cell wall, vacuole, cytoplasm, nucleus, nucleolus, and the position of the cell membrane.

Onion cell
(label structures)

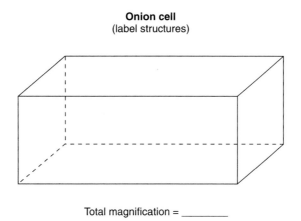

Total magnification = _____

Elodea

The aquatic plant *Elodea* provides another good example of plant cells. Use forceps to pluck a young leaf from the tip of a sprig of *Elodea* and place it on a slide with a drop of water and a coverslip. Examine the leaf under the microscope. Begin with low power and switch to high power to see detail. The leaf is two layers of cells thick. Use the fine adjustment knob to focus up and down with your microscope so that you can see the two layers. Examine the cells under high power. You should be able to see the following structures: (1) **cell wall**, (2) **vacuole**, and many small green (3) **chloroplasts** in the (4) **cytoplasm**. There is also a nucleus, but it is difficult to see among all the chloroplasts. You will also be able to see the cell membrane later. If you scan your slide you should be able to see some cells in which the chloroplasts are moving along the inside of the cell wall. It is actually the cytoplasm that is in motion; therefore, this phenomenon is known as cytoplasmic streaming.

Draw and label the three-dimensional structure of an *Elodea* cell in the space provided. Label the cell wall, cytoplasm, cell membrane, chloroplasts, and vacuole.

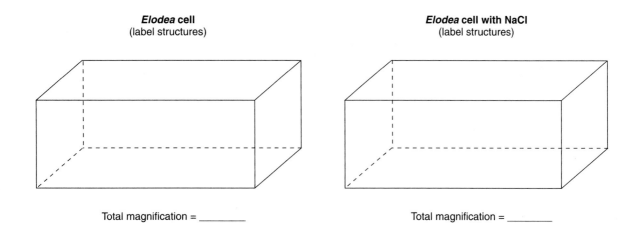

Elodea cell
(label structures)

Total magnification = _____

Elodea cell with NaCl
(label structures)

Total magnification = _____

After you have examined the living cells of the *Elodea* leaf, gently remove the coverslip, add a drop of 5% salt solution to the slide, and replace the coverslip. The salt solution will cause water to leave the large central vacuole, and the cell will shrink and pull away from the cell wall. This will allow you to see the **cell membrane**, which is on the outside of the cytoplasm. In the space provided, draw and label an *Elodea* cell after salt water is added.

Kingdom Protista

The kingdom **Protista** contains many different kinds of organisms in which each cell functions as a separate unit. The many kinds of Protista are lumped together in one kingdom for convenience and are subdivided into two major types of organisms, **algae** and **protozoa**. Algae are either single cells or groups of similar cells that have cell walls and are capable of photosynthesis. Protozoa lack cell walls and with a few exceptions are not capable of photosynthesis.

Spirogyra

The freshwater organism *Spirogyra* is a good example of an alga. *Spirogyra* is composed of cells that are attached to one another end-to-end to form long, hairlike strands. Use an eyedropper to obtain a few of these strands from the culture provided and prepare a slide for examination. In many ways a *Spirogyra* cell will appear similar to those of plants. It has a (1) **cell wall**, (2) **cytoplasm**, (3) a large **vacuole** between the strands of cytoplasm, and one or two spiral-shaped (4) **chloroplasts**. On the chloroplast you will be able to see dots. These are (5) **pyrenoids** which are places in the chloroplast where starch is manufactured. The cell also has a centrally located nucleus which is suspended in the center of the cell by strands of cytoplasm, but this is difficult to see without special stains. The cell also has a different shape from that of the plants you looked at previously. These cells are cylindrical rather than boxlike.

Draw and label a *Spirogyra* cell in the space below.

Euglena

Another member of the kingdom Protista is *Euglena*. Obtain a drop of the *Euglena* culture and place it on a slide with a coverslip. Examine it under the microscope. You should be able to see some organisms swimming around. *Euglena* has a long (1) **flagellum** at the anterior end that whips about and pulls the cell through the water. You will be able to see that they also have a red (2) **eyespot** at the base of the flagellum. Within the cytoplasm of the cell you will be able to see several green (3) **chloroplasts**. The eyespot allows the *Euglena* to swim toward a source of light and, therefore, position itself so that its chloroplasts receive sunlight. *Euglena* lacks a cell wall. Its outer covering is a flexible (4) **cell membrane**, so it is able to bend as it swims. It has a nucleus, but this is often difficult to see without special staining. Because it has chloroplasts, it is able to carry on photosynthesis; however, it is also able to "eat" by taking up organic molecules from its surroundings. Because it has chloroplasts, some people classify it with the algae. Because it swims, lacks a cell wall, and eats, some people prefer to classify it with the protozoa.

Paramecium

Paramecium is a large protozoan, which is just visible to the naked eye. Obtain a drop of culture medium containing *Paramecium*. They will have been fed yeast cells that were stained with the dye, congo red. You may need to place the organisms in a special syruplike, methyl cellulose solution to slow them down so that you can see them. Your instructor will provide the solution if needed. On their surface protruding through the (1) **cell membrane** are hundreds of tiny hairlike (2) **cilia** that they use for movement. On its surface you should be able to see a funnel-like structure through which the *Paramecium* feeds. Inside the organism you will see a number of spherical (3) **food vacuoles** containing yeast. Food vacuoles that were recently formed as the organism fed on the yeast will be red. Older food vacuoles in which digestion has begun will turn blue. Although the *Paramecium* does not have a cell wall, it does have a stiff outer layer of its cytoplasm. At either end of the cell you should be able to see a (4) **contractile vacuole** which periodically fills with water and collapses expelling water from the cell. When the contractile vacuoles are empty they will appear star-shaped. They become large spherical clear areas as they fill with water. *Paramecium* has a large **macronucleus** and one or more smaller **micronuclei**, but these are often difficult to see without staining.

Draw and label cells of *Euglena* and *Paramecium* in the space below.

Kingdom Animalia

Animal cells are often difficult to study because they are small. They are also easily destroyed when making a slide because they do not have a protective cell wall.

Cheek Epithelial Cells

Whenever human tissue is used in lab, we must follow special precautions. Therefore, the toothpick and the slides and coverslips you use must go in the special disposal container provided.

One kind of animal cell that is relatively easy to study is the cheek epithelial cells from the inside of your mouth. Take a toothpick and scrape the inside of your cheek (figure 30.3).

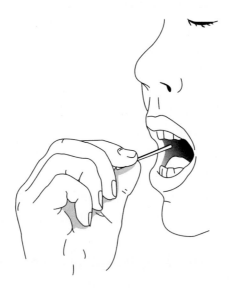

Gently scrape the inside of your cheek with
the broad end of a toothpick.

Figure 30.3

Smear the material from the toothpick onto a slide and add a drop of methylene blue stain and a coverslip. Use low power to locate some cells, then examine them under high power. You should be able to see flattened cells that have an irregular outline. (They will look like blue fried eggs.) The outside surface of the cell is the (1) **cell membrane**. You should also be able to see a football-shaped (2) **nucleus** in the (3) **cytoplasm** of the cell. On the surface of the cell you will be able to see a large number of tiny dots. These are (4) **bacteria**.

Draw a cheek epithelial cell in the space at the top of the next page.

Kingdom Fungi

Fungi are composed of cells that are attached end-to-end to form long filaments known as (1) **hyphae** (singular, **hypha**). These hyphae may form masses with no particular shape or may be organized to form very specific shapes such as mushrooms. We will look at the cells of a common mold. Obtain a sample of the mold and transfer it to a slide with a toothpick. Add a drop of methylene blue stain and a coverslip. Examine it under the microscope.

Although the cells are small, you should be able to distinguish the following structures: (1) a **cell wall** surrounding the cell (2) and large, clear **vacuoles** within the (3) **cytoplasm** of the cells. **Nuclei** are present but are very small and difficult to see without extremely high magnification. Some fungi have the hyphae divided into individual cells and usually have one nucleus per cell. Other fungi may have two nuclei per cell. Still other fungi do not have cross walls separating the hyphae into individual cells, and each hypha has several nuclei in it.

You will probably see a large number of spherical structures. These are reproductive structures called spores. The large number of spores produced and their small size makes them ideal mechanisms for distributing the mold to new sources of food.

Draw and label a fungal cell in the space below.

Domain Eubacteria

Most Eubacteria are extremely tiny and difficult to see. Because they lack a nucleus and most other kinds of organelles, it is difficult to see anything other than the general shape and size of the cells. You have already seen some bacteria on your cheek epithelial cells. Now you will view a mixture of bacteria cultured from soil and an example of cyanobacteria (blue-green algae), which is easily collected and viewed.

Soil Bacteria

Obtain a drop of the culture of soil bacteria, make a wet-mount slide, and examine under low power. What you see will be a mixture of many different kinds of organisms: protozoa, worms, algae, etc. Use high power to look for the tiny bacteria. Some of the largest soil bacteria will be corkscrew-shaped and will be swimming. These organisms do not have a nucleus and lack the other cellular structures typical of eukaryotic organisms. Therefore, it will be impossible to see any characteristics other than the general shape of the cells.

Cyanobacteria

Obtain a drop of the culture of *Anabaena*. It consists of strings of cells attached end-to-end like beads. These cells have a (1) **cell wall**. Often you will be able to see some larger cells that are specialized to withstand harsh environmental conditions. These specialized cells are called (2) **heterocysts**. They typically form when the algae begins to dry up from lack of water or when there is a significant change in the temperature. Few other structures are identifiable.

Draw and label examples of the bacteria and *Anabaena* in the space below.

Results

1. List two structural differences between prokaryotic cells and eukaryotic cells.

2. List two structural differences between plant and animal cells. How are these structural differences related to the ways the cells function?

3. In what ways do fungi resemble plant cells? In what ways are they different from plant cells?

4. Describe three ways in which algal cells and plant cells are similar.

5. Why are algae and protozoa placed in the same kingdom, Protista?

6. Describe the size and location of the vacuole in the onion cell. What does the vacuole contain?

7. Was the purpose of this lab accomplished? Why or why not? (Your answer to this question should show thoughtful analysis and careful, thorough thinking.)

Experiment 31: Enzymes

Invitation to Inquiry

Enzymes are always a part of all living things, even your foods. For this investigation you will need a package of gelatin desert, four clear glass containers, a carrot, some fresh pineapple, and miniature marshmallows.

Mix the gelatin according to directions and pour into four containers.

1. To the first container do not add anything. It will be your control.
2. Into the second container mix a quantity of shredded carrot.
3. Into the third container mix a quantity of crushed fresh pineapple.
4. Into the fourth container mix a quantity of miniature marshmallows.
5. Place all four containers in the refrigerator for the time specified on the gelatin dessert package.
6. When the proper time has passed, remove the four containers and examine their contents.
7. Describe what you see.
8. Write an explanation of what you think might be going on here.

Keep in mind that all living things carry out their metabolic reactions using enzymes. These enzymes are released when cells are damaged or destroyed, such as in cutting. Also recall that gelatin is a protein that is converted from a sol to gel state when it is cooled.

Background

Every living organism carries out a large number of chemical reactions. It is essential for the life of the organism that these reactions occur at an extremely rapid rate and at a safe temperature. All organisms contain **enzymes**, which are protein molecules that speed up the rate of chemical reactions without increasing the temperature.

All chemical reactions require an initial input of energy to get them started. This is called **activation energy**. Enzymes do not start a reaction; they merely speed up the reaction already in progress by reducing the need for large amounts of activation energy. Life on Earth would not be possible without this increased rate of reaction. Each reaction in a cell requires a specific enzyme to allow the reaction to proceed at the proper rate. Because there are hundreds of different reactions necessary in the life of the cell, hundreds of different enzymes are present in the cell. For an enzyme to work in a reaction, it must fit with its **substrate** (the molecule that will be altered). Each type of enzyme has a specific physical shape that fits the physical shape of its substrate.

When an enzyme reacts with a substrate, the two molecules physically combine to form an **enzyme-substrate complex**. The substrate is changed into the new **end product**, but the enzyme is not changed by the reaction. The number of times one molecule of enzyme can react with a substrate in a period of time is known as the **turnover number** (e.g., 500,000 per second). This means that one enzyme molecule reacts with 500,000 substrate molecules in a second.

If a molecule has a shape that is almost the same as that of the normal substrate, this nonsubstrate molecule might bind to the enzyme. When the enzyme is attached to this molecule, the enzyme is not free to combine with the normal substrate molecule. The nonsubstrate molecule attached to the enzyme is called a *competitive inhibitor*. An **inhibitor** slows down the normal turnover number of an enzyme because it does not allow the normal substrate to have access to the enzyme. A general equation for an enzymatic reaction is shown in figure 31.1.

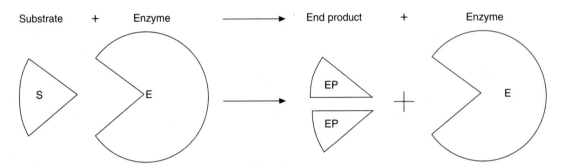

The specific reaction for today's exercise is:

Substrate		+ Oxygen	+ Enzyme	→	End product	+ Water	+ Enzyme
Pyrocatechol (colorless)		+ O$_2$	+ Tyrosinase	→	Hydroxyquinone (yellow-brown)	+ H$_2$O	+ Tyrosinase

Figure 31.1

In this exercise you are asked to determine how various factors influence the turnover number of an enzyme. The substrate is a clear, colorless molecule known as *pyrocatechol*. The enzyme is one naturally found in potatoes called *tyrosinase*. (You may use ground-up potatoes as an inexpensive source of the enzyme.) A reaction between this enzyme and the substrate results in the formation of an end product having a yellowish-brown color. The appearance of this color indicates that a reaction between the enzyme and the substrate has occurred. The degree of color change corresponds to the amount of end product produced. When performing this experiment, use the following scale to rank the degree of color change.

No change	Light yellow	Medium yellow	Golden brown
0	+	++	+++

Note: Potatoes are an inexpensive source of the enzyme tyrosinase. The enzyme is obtained by grinding a potato in water and straining the blended potato through a strainer. If you are using potatoes as a source of enzymes you will have best results if you do the following:

1. Grind up the potato just before you need it.
2. Use a chilled potato from a refrigerator and ice cold water to blend the potato.
3. After filtering the ground-up potato, store the liquid containing the enzyme in a small container in a beaker of ice.

If you use potato juice as a source of enzyme, you may see a pink color develop after a few minutes. This color change is not related to the reaction we are using in this exercise. Ignore any pink color that develops and record only the intensity of the yellow color that develops. Your instructor may use a pure solution of the enzyme tyrosinase. If this is used, no pink color appears.

During this lab exercise, you will observe the

1. normal reaction of the enzyme, value of a control in an experiment, and enzyme specificity.
2. influence of temperature on the turnover number.
3. influence of concentration of the enzyme on the turnover number.
4. influence of concentration of the substrate on the turnover number.
5. influence of pH on the turnover number.
6. influence of inhibitors on the turnover number.

When everyone has completed this exercise, your instructor may wish to discuss the results obtained by the class.

Procedure

Throughout this exercise it is important to clearly label all test tubes so that you can identify them.

Control

This portion of the exercise demonstrates the normal reaction that occurs between the enzyme (tyrosinase) and the substrate (pyrocatechol) used in today's exercise. You can also determine the value of a control and whether an enzyme is substrate specific. **Note**: *All glassware must be clean for this experiment to work properly.*

1. Put 10 ml of distilled water into a test tube and add 10 drops of enzyme. To this mixture add 10 drops of substrate. Mix the contents of the tube by holding the tube in one hand near the top and gently tapping the base of the tube with your other hand. Do not cover the opening of the test tube with your thumb!
2. Put 10 ml of distilled water into a second test tube. Add 10 drops of substrate and mix. This tube contains no enzyme.
3. Put 10 ml of distilled water into a third test tube. Add 10 drops of enzyme and 10 drops of sucrose (the wrong substrate). Mix this tube.

4. Observe each tube after 5 minutes and note any changes in color. Record the color in table 31.1. The color change in the first test tube indicates what normally occurs in this particular enzyme-substrate reaction. A control, a basis of comparison for a reaction, is an essential part of any experiment. This first test tube serves as a control during the remainder of this exercise. You can compare experimental tubes with this one to determine whether a reaction has occurred.

    a. What is the purpose of the second tube containing only water and substrate?

    b. What can you learn from the third tube containing water, enzyme, and sucrose?

Table 31.1	Enzyme Reactions			
Tube	Contents of tube	Original color	Color after 5 minutes	Rank
a	Water, 10 drops of enzyme, 10 drops of pyrocatechol	Clear		
b	Water, 10 drops of pyrocatechol	Clear		
c	Water, 10 drops of enzyme, 10 drops of sucrose	Clear		

Temperature

    In this portion of the exercise, you will examine the effect of different temperatures on the activity of the enzyme. You need to allow the water in the test tubes to come to a specific temperature before you add the substrate and enzyme. After these are mixed with the water, they need to stay at that specific temperature for 5 minutes to see if the temperature influences the enzyme's activity.

1. Take five test tubes and label them a, b, c, d, and e. Fill each with 10 ml of distilled water. Add 10 drops of enzyme to each tube.
2. Place tube a in an ice water mixture in a water bath for 5 minutes to allow it to cool. Similarly, place tubes b through e in water baths of 20°, 40°, 60°, and 100°C, respectively. These tubes are to remain in their respective water baths until they reach the designated temperature.
3. After allowing the tubes to come to their appropriate temperature, add 10 drops of substrate to each tube. Mix the contents and immediately return the tubes to their appropriate water baths for an additional 5 minutes.
4. After 5 minutes, remove the test tubes from the water baths and observe the color of the tubes. Record the color intensity of each tube in table 31.2 and rank them from darkest to lightest. Graph your results on the grid provided.

Table 31.2	Temperature			
Tube	Contents of tube	Original color	Color after 5 minutes	Rank
a 0°C	Water, enzyme, and substrate	Clear		
b 20°C	Water, enzyme, and substrate	Clear		
c 40°C	Water, enzyme, and substrate	Clear		
d 60°C	Water, enzyme, and substrate	Clear		
e 100°C	Water, enzyme, and substrate	Clear		

Effect of Temperature on Amount of End Product

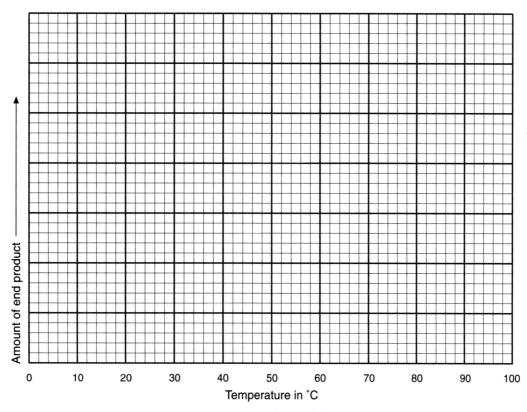

Amount of end product

Temperature in ˚C

Concentration of Enzyme

In this experiment you will examine the effect of altering the number of enzyme molecules on the number of product molecules produced.

1. Take three test tubes and label them *a*, *b*, and *c*. Fill each with 10 ml of distilled water. Place them in a test tube rack.

2. Add three drops of enzyme to tube *a*, nine drops of enzyme to tube *b*, and 27 drops of enzyme to tube *c*.

3. Add 10 drops of substrate to each of these tubes. Mix thoroughly and observe after 5 minutes.

4. Record the color intensity of each tube in table 31.3 and rank them from darkest to lightest. Graph your results on the grid provided.

Table 31.3	Concentration of Enzyme			
Tube	Contents of tube	Original color	Color after 5 minutes	Rank
a	Water, 3 drops of enzyme, and 10 drops of substrate	Clear		
b	Water, 9 drops of enzyme, and 10 drops of substrate	Clear		
c	Water, 27 drops of enzyme, and 10 drops of substrate	Clear		

Effect of Enzyme Concentration on Amount of End Product

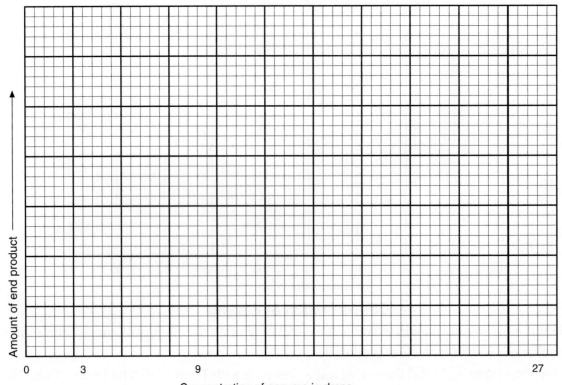

Amount of end product

0   3   9   27

Concentration of enzyme in drops

Concentration of Substrate

In this experiment you will examine the effect of altering the number of substrate molecules on the number of product molecules produced.

1. Take three test tubes and label them *a*, *b*, and *c*. Fill each with 10 ml of distilled water. Place them in a test tube rack.
2. Add three drops of substrate to tube *a*, nine drops of substrate to tube *b*, and 27 drops of substrate to tube *c*.
3. Add 10 drops of enzyme to each of these tubes. Mix thoroughly and observe after 5 minutes.
4. Record the color intensity of each tube in table 31.4 and rank them from darkest to lightest.

Table 31.4	Concentration of Substrate			
Tube	Contents of tube	Original color	Color after 5 minutes	Rank
a	Water, 10 drops of enzyme, and 3 drops of substrate	Clear		
b	Water, 10 drops of enzyme, and 9 drops of substrate	Clear		
c	Water, 10 drops of enzyme, and 27 drops of substrate	Clear		

Effect of Substrate Concentration on Amount of End Product

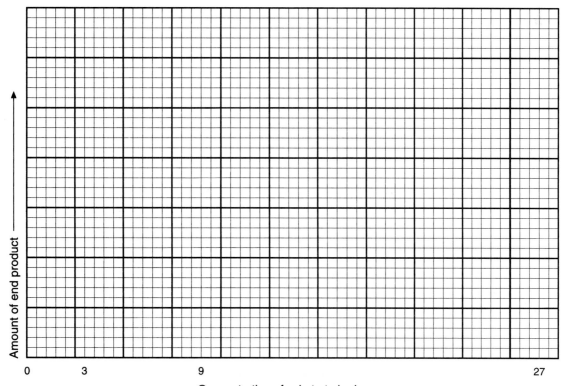

Concentration of substrate in drops

253

pH

pH is a measure of the number of hydrogen ions (H$^+$) present in a solution. Solutions that have many hydrogen ions are called acids and have a low pH. Solutions with few hydrogen ions are called bases and have a high pH. Solutions with a pH = 7 are called neutral solutions. In this experiment you will assess the effect of different pH values on the effectiveness of enzymes.

1. Take five test tubes and label them *a*, *b*, *c*, *d*, and *e*. Fill tube *a* with 10 ml of water that has been adjusted to a pH of 3. Place the tube in a test tube rack.

2. Fill tube *b* with 10 ml of water that has been adjusted to a pH of 5. Place the tube in a test tube rack.

3. Fill tube *c* with 10 ml of water that has been adjusted to a pH of 7. Place the tube in a test tube rack.

4. Fill tube *d* with 10 ml of water that has been adjusted to a pH of 9. Place the tube in a test tube rack.

5. Fill tube *e* with 10 ml of water that has been adjusted to a pH of 11. Place the tube in a test tube rack.

6. Add 10 drops of enzyme to each tube.

7. Add 10 drops of substrate to each tube. Mix the tubes thoroughly and observe after 5 minutes.

8. Record the color intensity of each tube in table 31.5, and rank them from darkest to lightest. Graph your results on the grid provided.

Table 31.5 Effect of pH

Tube	Contents of tube	Original color	Color after 5 minutes	Rank
a	Water adjusted to pH of 3, 10 drops of enzyme, and 10 drops of substrate	Clear		
b	Water adjusted to pH of 5, 10 drops of enzyme, and 10 drops of substrate	Clear		
c	Water adjusted to pH of 7, 10 drops of enzyme, and 10 drops of substrate	Clear		
d	Water adjusted to pH of 9, 10 drops of enzyme, and 10 drops of substrate	Clear		
e	Water adjusted to pH of 11, 10 drops of enzyme, and 10 drops of substrate	Clear		

Effect of pH on Amount of End Product

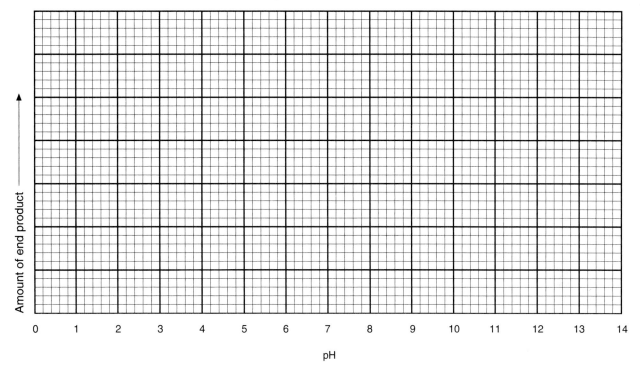

pH

(y-axis) Amount of end product

Inhibitors

Inhibitors interfere with the ability of enzymes to interact with enzymes. In this experiment you will identify one of two molecules as being an inhibitor. (The following steps must be performed in the order presented or results will vary.)

1. Take six test tubes and label them *a* through *f*. Fill each tube with 10 ml of distilled water. Place them in a test tube rack.
2. Add one drop of phenylthiourea to tube *a*, 10 drops of phenylthiourea to tube *b*, and 20 drops of phenylthiourea to tube *c*.
3. Add one drop of tyrosine to tube *d*, 10 drops of tyrosine to tube *e*, and 20 drops of tyrosine to tube *f*.
4. Add 10 drops of enzyme to each of the six tubes.
5. Add 10 drops of substrate to each of the six tubes. Mix the tubes thoroughly and observe after 5 minutes.
6. Record the color intensity of each tube in table 31.6 and rank them from darkest to lightest.

Tube	Contents of tube	Original color	Color after 5 minutes	Rank
a	Water, 1 drop of phenylthiourea, 10 drops of enzyme, and 10 drops of substrate	Clear		
b	Water, 10 drops of phenylthiourea, 10 drops of enzyme, and 10 drops of substrate	Clear		
c	Water, 20 drops of phenylthiourea, 10 drops of enzyme, and 10 drops of substrate	Clear		
d	Water, 1 drop of tyrosine, 10 drops of enzyme, and 10 drops of substrate	Clear		
e	Water, 10 drops of tyrosine, 10 drops of enzyme, and 10 drops of substrate	Clear		
f	Water, 20 drops of tyrosine, 10 drops of enzyme, and 10 drops of substrate	Clear		

Table 31.6   Inhibitors

Results

1. Which of the two substances (phenylthiourea or tyrosine) was the inhibitor in the exercise you just completed? What was your evidence?

2. If you were to look at the three test tubes involved in the enzyme concentration experiment after 24 hours, how would the colors of each compare? Explain.

3. Both high and low temperatures reduced the amount of color (product) produced. However, the cause of the reduction is different in the two cases. Explain what the differences are.

4. Since an enzyme-substrate complex must be formed for an end product to be produced, how might an inhibitor reduce the effectiveness of an enzyme?

5. What is turnover number? How did you estimate turnover number in this series of experiments?

6. Why does increasing the amount of substrate increase the amount of color produced in the test tubes?

7. At what pH was the enzyme most effective?

At what other pHs did the enzyme work but not as effectively?

Explain why changing the pH alters the effectiveness of the enzyme.

8. Was the purpose of this lab accomplished? Why or why not? (Your answer to this question should show thoughtful analysis and careful, thorough thinking.)

Experiment 32: Photosynthesis and Respiration

Invitation to Inquiry

Try this.
1. Get an aquatic plant from a pond or pet store.
2. Place one of the plants in a clear glass container of tap water.
3. Fill another clear glass container with tap water.
4. Place both containers in a well-lit area such as under a fluorescent light.
5. Observe both carefully.
6. Count the bubbles on the interior glass surface of both containers.

Collect and record your data. Do you see a difference in the number of bubbles in the two containers? Hypothesize what the bubbles might be. Why might there be a difference?

Background

Photosynthesis is a metabolic process that combines carbon dioxide (CO_2) and water (H_2O) to form sugar ($C_6H_{12}O_6$) and oxygen (O_2). The process takes place within the chloroplasts of plants and algae. The chloroplasts contain the green pigment **chlorophyll** and the enzymes necessary for photosynthesis. The chlorophyll traps light that serves as the energy source that allows the process to take place. A simplified equation for photosynthesis is

$$6\,H_2O + 6\,CO_2 \xrightarrow[\text{Chloroplasts}]{\text{Light}} C_6H_{12}O_6 + 6\,O_2\,.$$

$$\underbrace{}_{\text{Inorganic raw materials}} \qquad \underbrace{\phantom{C_6H_{12}O_6 + 6\,O_2}}_{\text{End products}}$$

All organisms require energy to sustain themselves. Nearly all organisms, including plants and animals, carry on **aerobic respiration** in which sugar and oxygen react to form carbon dioxide, water, and a source of energy known as ATP (adenosine triphosphate). Mitochondria are cellular structures that contain the enzymes necessary for the many individual steps of aerobic respiration. A simplified equation for aerobic respiration is

$$C_6H_{12}O_6 + 6\,O_2 \xrightarrow[\text{Mitochondria}]{\text{Enzymes}} 6\,H_2O + 6\,CO_2 + \text{Energy}\,.$$

$$\underbrace{\phantom{C_6H_{12}O_6 + 6\,O_2}}_{\text{Raw Materials}} \qquad \underbrace{}_{\text{End Products}}$$

Look closely at the balanced equations for photosynthesis and respiration and notice that the end products of one reaction are the raw materials for the other. Only organisms containing chlorophyll can perform photosynthesis, whereas respiration can take place in virtually every organism.

Many plants perform photosynthesis and respiration simultaneously. The **P/R ratio** (photosynthesis/respiration ratio) compares the rate of photosynthesis to the rate of respiration. Knowing this ratio can help explain what happens in a plant at different times in its life. For instance, the P/R ratio is different for a corn plant during spring, summer, and fall. It is also different for day and night. Animals, on the other hand, can engage only in respiration.

Because animals are incapable of converting inorganic raw materials into organic molecules, they must obtain energy-rich organic molecules by eating plants or other animals. They also need oxygen to allow them to release energy from organic molecules. Thus, animals are dependent on plants for the two end products of photosynthesis, namely organic molecules (glucose) and oxygen.

Water has many gases dissolved in it, including oxygen and carbon dioxide. Aquatic organisms use these gases when they carry on photosynthesis and aerobic respiration and release gases into the water as well. This exercise allows you to make measurements of the rates of photosynthesis and respiration by measuring the amount of oxygen and carbon dioxide present in the water in which the organisms live.

This laboratory activity gives you an opportunity to set up an experiment with proper controls, quantitatively test water samples for oxygen and carbon dioxide content, and collect and analyze data. You may also learn that experimental work sometimes yields results that are difficult to interpret.

During this exercise you allow organisms to engage in their normal biochemical processes. Evidence that the organisms have carried on photosynthesis or respiration is revealed by sampling the oxygen and carbon dioxide content of the water in which they live. The tests for measuring the oxygen and carbon dioxide content appear on pages 262 and 264. Follow these test procedures carefully, because you will be measuring very small quantities—**parts per million (ppm)**—of oxygen and carbon dioxide. If your water sample contains 8 ppm of oxygen, it means that there are eight oxygen molecules dissolved in every 1 million molecules of your sample.
During this lab exercise, your group will

1. determine dissolved oxygen and dissolved carbon dioxide concentration in aged tap water. (The purpose for doing this is to get baseline data as well as to give you practice with these complex tests.)

2. set up controls of aged water and three experimental situations: plants in light, plants in darkness, and fish.

3. determine dissolved oxygen and dissolved carbon dioxide concentrations of the controls and the three experimental situations at the end of an hour.

4. use the data collected to answer questions.

Procedure

Initial Trial

Fill a large beaker from the container labeled aged water. The aged water is simply tap water that has been sitting open to the atmosphere overnight. Therefore, the aged water has the same concentrations of carbon dioxide and oxygen dissolved throughout the container. In addition, it is equilibrated to room temperature. Since everyone in class will use this aged water to set up their experiments, everyone will start with water that contains the same amount of carbon dioxide and oxygen and has the same temperature. Test this aged water for dissolved O_2 and CO_2. The directions for the dissolved oxygen test and the dissolved carbon dioxide test are found on pages 263 and 265. Record the results in the proper column of Data Table 32.1.

Data Table 32.1	Dissolved Oxygen and Carbon Dioxide Results	
	Dissolved O_2 (ppm) (number of drops of sodium thiosulfate)	Dissolved CO_2 (ppm) (number of drops of NaOH)
Initial aged water		
Control in light		
Control in dark		
Plant in light		
Plant in dark		
Goldfish		

Controls

Fill four large test tubes with aged water; cork the tubes in such a way that no air is trapped. Label these tubes Controls and place two of them in a test tube rack marked Light. Place the other two control tubes in a test tube rack in the dark. Your instructor will designate this location. At the end of the hour you will use the water from one of each pair of tubes to test for oxygen content and the water in the other tube to test for carbon dioxide content.

Experimental Tubes

Plant in Light

Fill two other large tubes with aged water. Place several healthy, green sprigs of *Elodea* or other water plant in each tube. There should be plants from the top to the bottom of the test tube, but the plants should not be jammed together in a clump. Cork the tubes without trapping air. Label these tubes Plant in Light and place them in the test tube rack in front of a fluorescent light source for 1 hour. It is best to use fluorescent lights because incandescent lights tend to heat up the water in the tubes and change the amount of gases that can remain dissolved in the water. See table 32.1.

Plant in Darkness

Fill two more large tubes with aged water and *Elodea* or other water plant; cork tubes. Label these tubes Plant in Dark and place them in the designated dark area for 1 hour.

Table 32.1   Effect of Temperature on Water Solubility of O_2 and CO_2

Temperature °C	O_2 Solubility mg/kg or ppm	CO_2 Solubility mg/kg or ppm
0	14.62	3347
5	12.77	2782
10	11.29	2319
15	10.08	1979
20	9.09	1689
25	8.26	1430
30	7.56	1250
35	6.95	1106
40	6.41	970

Dissolved Oxygen Test

Follow this procedure to determine the dissolved oxygen content of the initial aged water, control containers, and the three sets of experimental containers.

1. Carefully fill (overflow) the small glass-stoppered bottle with the water to be tested. (Do not put plants or fish into this bottle). (Do not use water that has been used for any other test.) Insert the glass stopper and pour off any excess water trapped on the outside of the bottle around the stopper.
2. Remove the stopper. Cut open chemical packets *a* and *b* and carefully empty the contents of both into the bottle. Since the quantity of chemicals in the packets is premeasured, it is important that you not spill any of the contents and that you get all the contents into the bottle. Replace the stopper carefully (do not trap any air bubbles) and shake for 30 seconds or until all granules are dissolved. A brown precipitate will form. (If granules still remain after 30 seconds of shaking, proceed to step 3.)
3. Set the bottle on the lab table. After the brown precipitate has settled out about halfway, empty the contents of the large packet *c* into the bottle. Stopper and shake again until the precipitate has completely dissolved. If you have brown particles in the water, vigorous shaking should get rid of them. You should now have a clear, yellowish liquid.
4. Fill the small measuring test tube with the yellow liquid and pour one full tube of the yellow liquid into a clean glass beaker. It is a good idea to rinse any glassware before you use it to be sure that it is clean.
5. While gently swirling the beaker to mix its contents, add sodium thiosulfate solution drop by drop (count the drops) until the yellow color turns clear, like water. The color change is best seen when the beaker is placed over white paper. The number of drops of sodium thiosulfate used equals the parts per million (ppm) of dissolved oxygen. Your result should be between 1 and 20 ppm. If your result is different from this, call it to your instructor's attention. Your instructor should be able to help you determine what is wrong.
6. When you are finished with the test, pour the liquids into the large waste chemicals container provided and rinse your glassware.

Animal

Fill two other large tubes with aged water and place a goldfish in each test tube. Cork the tube without trapping air bubbles. Label these tubes Goldfish and place them with the plant in the light for 1 hour.

Analysis of Results

Often a visual presentation of data helps one see patterns or trends and makes interpretation easier. Use the data from Table 32.1 to construct a bar graph in the space provided below. Use two different colors or shades to distinguish between oxygen and carbon dioxide.

263

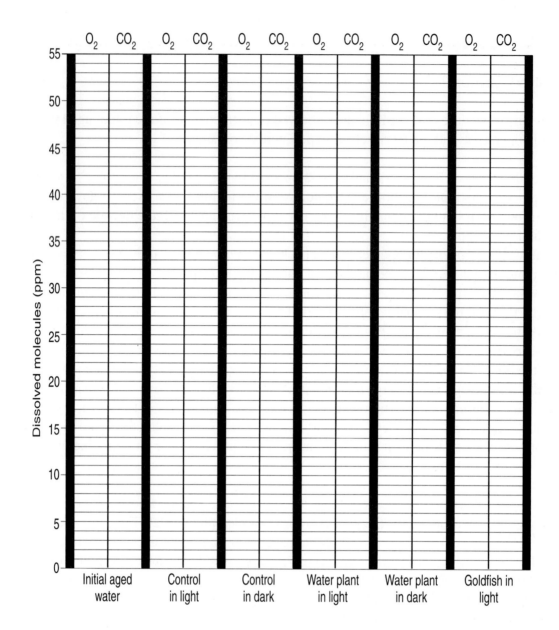

Results

Use the equations for photosynthesis and aerobic respiration to help you think about the questions. Remember reactants decrease in concentration as products increase in concentration.

Photosynthesis:

$$6\,H_2O + 6\,CO_2 \xrightarrow[\text{Chloroplasts}]{\text{Light}} C_6H_{12}O_6 + 6\,O_2$$

Aerobic Respiration:

$$C_6H_{12}O_6 + 6\,O_2 \xrightarrow[\text{Mitochondria}]{\text{Enzymes}} 6\,H_2O + 6\,CO_2 + \text{Energy (ATP)}$$

264

Dissolved Carbon Dioxide Test

Follow this procedure to determine the dissolved carbon dioxide content of the initial aged water, control containers, and the three sets of experimental containers.

1. Use a graduated cylinder to measure 30 ml of water to be tested. (Do not put plants or fish into the bottle). (Do not use water that has been used for any other test.) Pour the 30 ml of water into a small beaker.
2. Add five drops of phenolphthalein to the water. (Phenolphthalein is an acid/base indicator.)
3. While gently swirling the beaker to mix its contents, add sodium hydroxide solution drop by drop (count the drops) until a light pink color appears and stays pink. The change to a pink color is easiest to detect when the beaker is placed over white paper. The number of drops of sodium hydroxide solution used to get the pink color equals the parts per million (ppm) of dissolved CO_2 in the water. Your result will be between 1 and 20 ppm. If your results are different from this, call it to your instructor's attention. Your instructor should be able to help you determine what is wrong.
4. When you are finished with the test, pour the liquids into the large waste chemicals container provided and rinse your glassware.

1. How does the P/R ratio change from summer to winter for a plant growing in Canada?

2. What does it mean if the P/R ratio is 2/1? Which of the tubes (Plant in light, Plant in dark, goldfish) had a P/R ratio similar to 2/1?

3. Why was there less CO_2 in the Water Plant in Light tube at the end of the hour than in the control?

4. If you could have measured the number of individual water molecules, would the number of water molecules in the tube have differed at the beginning and at the end of the hour for the Water Plant in Light? Why?

5. Why was there less oxygen in the Water Plant in Dark tube at the end of the hour?

6. Why was there less oxygen in the Goldfish tube at the end of the hour?

7. How do you think the results would have differed if the Goldfish tube had been placed in the dark?

8. Compare the processes occurring in the plants in the dark with the processes occurring in the goldfish in the light or in the dark.

9. If a fish had been placed in the tube with the plants in the light, what results would you expect? Why?

10. If a fish had been placed in the tube with the plants in the dark, what results would you expect? Why?

11. If you have $10,000 and you loaned your friend a nickel, how many parts per million of your money have you given to your friend?

Experiment 33: The Chemistry and Ecology of Yogurt Production

Invitation to Inquiry

Fermentation can result in either souring or spoiling. These are two different things! We eat many soured foods such as yogurt, cheese, and pickles. They do not cause illness. However, foods that have spoiled can be very harmful. One method of preventing spoilage is to add salt (NaCl) to the food.

1. Get two kinds of cottage cheese: one with salt and the other without salt.
2. Open each container and mix the cheese curds with your unwashed hands (preferable dirty hands!).
3. Place these containers in a warm place with the covers on top but not sealed.
4. Note the time.
5. Examine the containers every 6 hours and take notes on the changes that take place including such items as
    a. color.
    b. texture.
    c. smell.
6. Write a summary of what you have observed and draw conclusions about what took place.

Background

Yogurt is the product of the *fermentation* of milk sugar (lactose) by two different species of bacteria, *Lactobacillus bulgaricus* and *Streptococcus thermophilus*. **Fermentation** is a metabolic process in which organic molecules are broken down to simpler compounds by organisms <u>without the use of oxygen</u>. It is an example of **anaerobic respiration**. Organisms that carry on fermentation obtain energy in the form of ATP molecules. In the process of either aerobic or anaerobic respiration, each step in the process is controlled by an enzyme that converts a **substrate**, the material acted upon by the enzyme, into an **end product**. The end product of the first reaction becomes the substrate for the next reaction in the series. The **enzymes** in the bacteria we are using today convert **lactose** into **lactic acid**. Lactic acid and ATP are the final end products of this example of fermentation. A simplified version for fermentation is

$$\text{Glucose} \xrightarrow{\text{Enzymes}} 2 \text{ lactic acid } + 2 \text{ ATP} .$$

Figure 33.1 provides a more detailed description of what happens during fermentation.

Because acids have a sour taste, the result of the activity of these bacteria is a sour-tasting dairy product. The production of the acid is responsible for the change in the consistency of the milk. The greatly lowered pH causes the milk protein to coagulate and become thick and viscous. *Streptococcus thermophilus* also produces some other compounds that are important in determining

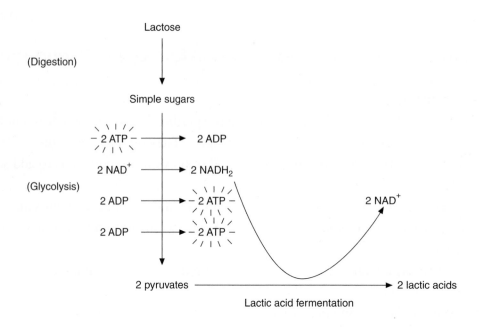

Figure 33.1

the final flavor of the yogurt.

To provide a consistently palatable product, commercial producers of yogurt must have a thorough understanding of the biology of the bacteria they use. They must be aware that changes in the environment alter the activities of the bacteria. Fresh milk normally contains a mixture of different species of bacteria that eventually ferment, or sour, the milk. Because these bacteria produce unwanted flavors and compete with the desired bacteria, it is desirable to reduce the natural bacterial population as much as possible. This is usually done by heating the milk for a specified period of time.

Substances that retard the growth of the desired species of bacteria—inhibitors—are another problem for the yogurt producer. These inhibitors are of two varieties. One type is produced naturally by the cow (it makes sense that cows would have evolved a method of inhibiting fermentation while the milk was still in the cow; otherwise, cows might give yogurt). The second kind of inhibitor is a much greater problem and results from the treatment of cows with antibiotics. If cows are being treated with penicillin or some other antibiotic that slows down the growth of disease-causing bacteria, this antibiotic can get into the milk and prevent the desired bacteria from doing their job of making yogurt.

With one carton of milk per student, you will

1. prepare and inoculate the container of milk with a bacterial culture that will produce yogurt.
2. incubate the mixture at the proper temperature.

Procedure

1. Empty a one-half-pint container of milk into a clean beaker and place on a hot plate.
2. Add to the milk 11 g of powdered milk. Stir and heat to 96°C. Do not allow the milk to boil. Stir constantly to avoid burning. *As you approach 96°C, stir constantly or the milk will boil over.*
3. As soon as the milk reaches 96°C, use hot pads to remove it from the hot plate. Allow the milk to cool to 46°C. Stirring occasionally will reduce the time needed to cool the milk.
4. Add a teaspoon of starter yogurt culture to your empty milk carton from the bulk culture provided.
5. Pour the cooled milk into the carton; close and staple shut. Label the carton with your name and the date.
6. Place your culture in the designated incubator (39°C) until it coagulates (about 6 to 8 hours). A high temperature of incubation results in a more sour yogurt; a lower temperature of incubation results in a more viscous product. The 39°C temperature is a compromise between these two temperatures. Your product should be sour and viscous like yogurt.
7. Remove the carton from the incubator following coagulation and cool to 10°C in a refrigerator.
8. When cooled (any convenient time within a week) note the odor and flavor. Compare this to a commercially prepared carton of yogurt.

Flavored Yogurts (Optional)

You may wish to add some flavoring to the yogurt. The flavorings are added after the incubation period to prevent their change in taste by the bacteria. Here are some suggestions for flavoring your yogurt.

cucumbers and black pepper	fruit preserves
dry fruit-flavored gelatins	nuts
honey	granola
chocolate syrup	cereals
fresh or frozen fruit	

Yogurt Cheese

Yogurt and cheese are both fermented milk products; however, cheeses have the curd (the solid portion) separated from the whey (the liquid portion). The curd of yogurt is also softer and more tart than most cheeses. To convert yogurt into a cheese, you merely need to allow the whey to separate from the curd so that the curd can form into a more firm product. This is easily done by placing the yogurt into a bag of closely knit cheesecloth and suspending the bag over a container to collect the whey. If you suspend the yogurt and allow the whey to drain away for 24 hours, the cheese that is formed has the consistency of cream cheese. If it is allowed to drain for a longer period, the curd becomes more dense and can be sliced. The whey can be saved. It is sometimes used in place of milk as a beverage or on cereals or as an additive in other recipes.

Results

1. Why was the milk heated prior to adding the starter culture?

2. What active ingredient(s) can be found in the yogurt starter culture?

3. What would happen if you added the yogurt culture to the heated milk before it had been allowed to cool?

4. What things might act as inhibitors in the yogurt-making process?

5. Now that you have made yogurt, what do you think is different about the process of making Swiss cheese, cheddar cheese, and yogurt cheese? (See directions for making yogurt cheese on page 269.)

6. What would happen if you added the yogurt culture to the cooled milk and immediately refrigerated it?

7. Which of the following terms apply to the process of making yogurt? (circle them)

aerobic	anaerobic
bacteria	yeast
carbon dioxide	methane
glycolysis	Krebs cycle
coenzyme	alcohol
lactic acid	fermentation
cellular respiration	

Experiment 34: DNA and RNA: Structure and Function

Invitation to Inquiry

All mammals produce milk as the primary food for their young. Milk contains many important nutrients including water, fat, protein, and the sugar, lactose. Normally, after weaning, mammals close down the enzyme systems responsible for the digestion of lactose in milk. Various kinds of research suggest that our human ancestors were similar to other mammals and lost their ability to digest milk as they grew and stopped drinking breast milk.

If adult mammals drink milk they are unable to digest the lactose in milk, and bacteria in their intestines ferment the lactose, resulting in the production of gases that produce symptoms such as distention of the belly, abdominal cramps, diarrhea, and foul-smelling stools. This condition is known as lactose intolerance. However, in many cultures the raising of cattle allows for a source of milk that can be consumed as food by adults, and approximately 50 percent of the human population is currently lactose tolerant. This means that they can digest the lactose in milk without experiencing the symptoms of lactose intolerance.

Recent research has indicated that lactose tolerance came about as a result of a mutation in the DNA of certain people in Eastern Europe. Diagram a scheme of events which would have started with such a mutation and resulted in the spread of this mutation throughout Europe and the world. Keep in mind that the lactose tolerance trait is primarily centered in the Caucasian population.

Background

It is to your advantage when doing this exercise to keep your textbook open to the chapter that describes nucleic acids and processes of replication, transcription, and translation. Refer to the information and diagrams in your text as needed.

Both of the nucleic acids, **deoxyribonucleic acid (DNA)** and **ribonucleic acid (RNA)**, are constructed of smaller subunits called **nucleotides**. The different kinds of nucleotides synthesized in the cell differ from one another in the kinds of sugars (deoxyribose or ribose) and the kind of nitrogen base (adenine, guanine, cytosine, thymine, or uracil) they contain. Cells maintain a supply or pool of these nucleotides for use in the processes of *DNA replication* and *RNA transcription*.

DNA replication is the series of chemical reactions that result in the formation of two identical double-stranded DNA molecules from one original molecule. This is an assembly process that uses the two strands of the existing DNA molecules as templates upon which new DNA nucleotides are aligned. **Transcription** involves the formation of a copy of the DNA code in the form of RNA molecules. The actual process of manufacturing the RNA is similar to replication, except that RNA nucleotides are matched to only one side of the DNA double helix. RNA molecules are single-stranded. There are three types of RNA produced: messenger RNA (mRNA), transfer RNA (tRNA), and ribosomal RNA (rRNA). Each of these RNA molecules has a special role to play in the functioning of a cell. The mRNA carries the message from the DNA in the nucleus to the cytoplasm of the cell. The tRNA carries amino acids used for the manufacture of proteins. The rRNA along with some proteins forms ribosomes, structures needed to assemble proteins. **Translation** involves the cooperation of all the kinds of RNA to bring about the synthesis of specific proteins. During translation the structure of the ribosome promotes complementary **base pairing** between mRNA and tRNA, leading to the formation of peptide bonds between specific amino acids. Strings of amino acids are known as **polypeptides**. One or more polypeptides are combined to form essential proteins such as enzymes; antibodies; hemoglobin; collagen, found in connective tissue; or myosin, found in muscle tissue.

This exercise will help you understand the details of replication, transcription, and translation. Figure 34.1 summarizes these central concepts of molecular genetics commonly called the central dogma.

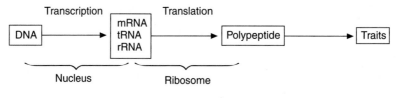

Figure 34.1

This exercise simulates the processes of DNA replication, RNA transcription, and translation. Remember, real DNA is a three-dimensional double helix. The flat plastic structures used in this exercise cannot be made to exactly duplicate the real molecules involved in these cell processes.

Work in pairs and be sure that you discuss with your partner what each stage of the exercise represents. If you are unsure at any time, check with the instructor.

During this lab exercise you will

1. separate the molecular models into pools.
2. use the models from the pools to construct DNA nucleotides.
3. use the DNA nucleotides to construct a model of a single strand of DNA.
4. use the DNA nucleotides to base-pair with the nucleotides on the single strand, thus forming a double strand of DNA.
5. replicate the double strand of DNA.
6. use the models from the pool to construct RNA nucleotides.
7. use the RNA nucleotides to construct tRNA molecules and attach the appropriate amino acid to these molecules.
8. transcribe DNA into mRNA.
9. translate and form a protein model.

Procedure (Don't Take Shortcuts!)

Building Double-Stranded DNA

1. *Separate the pieces of the molecular model into pools.* Each pair of students should get a DNA/RNA model kit. Empty the plastic parts onto the table and separate the various parts into piles (pools). These parts represent the types of molecules that are necessary to construct nucleotides. Table 34.1 lists the parts and the numbers of each that should be in your kit. If you find that you do not have the proper number, or if pieces are broken, check with your instructor.

Table 34.1	DNA-RNA Model Kit		
Part Name	Symbol	Number needed	Number present
Deoxyribose sugar	D	36	
Ribose sugar	R	18	
Phosphate	P	36	
Uracil	U	5	
Thymine	T	10	
Guanine	G	8	
Cytosine	C	8	
Adenine	A	10	
Hydroxyl group	OH	3	
Hydrogen atom	H	3	
Amino acids	Leu, His, Gly	1 of each type	

2. Construct DNA nucleotides. A DNA nucleotide is composed of a phosphate molecule bonded to a deoxyribose sugar molecule bonded to a nitrogen-containing base such as adenine. Use the models from the pools to construct DNA nucleotides.

<pre>
           Adenine   A
              | |
Phosphate — Deoxyribose   P — D
</pre>

a. Assemble 36 separate DNA nucleotides. Each nucleotide should look like one of the diagrams in figure 34.2.

b. Assemble

10 separate nucleotides using adenine,

8 separate nucleotides using guanine,

8 separate nucleotides using cytosine, and

10 separate nucleotides using thymine.

c. Make sure that all the phosphates are on the left side of the sugar. These 36 nucleotides make up the DNA nucleotide pool of your cell.

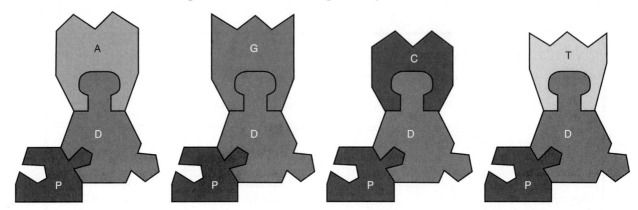

Figure 34.2 DNA nucleotides. Notice that each nucleotide differs in the type of nitrogen base it contains and not in the sugar or the phosphate.

3. Construct one side of a DNA molecule. Use the DNA nucleotides you have just assembled to construct a model of one side of a DNA helix using precisely the sequence given in figure 34.3. Attach the nucleotides from the pool, end-to-end, with the deoxyribose to the right and the phosphates to the left. The phosphate of one nucleotide should link with the sugar of the next. This sequence of nucleotides will represent the gene in our simulation. During the process of transcription, use this specific sequence as the template for building the mRNA molecule.

Figure 34.3 Single DNA strand

4. Construct a DNA double helix. Use complementary nucleotides to base-pair with the nucleotides on the single strand you just assembled to form a double-stranded DNA molecule. To assemble the second strand properly, follow the rule that the nitrogen base, adenine, always base-pairs with the base, thymine, in DNA and that the nitrogen base, guanine, always base-pairs with the base, cytosine. Slide the appropriate nucleotides from their pools into position so that a ladderlike molecule is formed on the table. Again, link the sugar of one nucleotide to the phosphate of the next nucleotide. The DNA is really a spiral molecule, but we will use this ladderlike molecule to represent the double-stranded DNA molecule.

276

Replication: Synthesis of DNA

When a cell divides, each of the two, new, daughter cells must receive the same DNA. To accomplish this, the cell usually has two sets of DNA molecules on chromosomes that may be separated into the two newly forming cells. The process of constructing copies of a double-stranded DNA molecule is called DNA replication. If the genetic information is not replicated and separated equally when cells divide, the new cells will not contain the genetic messages necessary to manufacture the proteins needed by the cell.

Replication of DNA is going on within your body at this moment. Whenever a wound is repaired, growth occurs or old cells are replaced, DNA must first replicate. In other words, before any cell division occurs, the DNA molecules make copies of themselves so that each new cell receives the same genes. The diagrams in figure 34.4 show a cell in the process of division after the DNA has replicated.

Each pair of chromatids contains identical DNA.

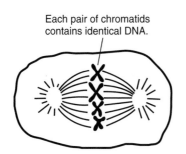

Chromatids separate and the identical DNA moves into two new daughter cells.

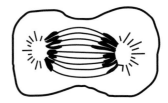

Figure 34.4

You are now ready to proceed through the process of DNA replication. The way in which you move the pieces of the model is similar to what actually happens in a cell during the replication process.

1. Separate the two strands of DNA.
2. Use the remaining DNA nucleotides from the pool and match the proper nucleotides from the pool with their partners on the two separated DNA strands (A- T, G-C).

When this is completed, you should have two double-stranded DNA molecules in front of you on the table. Furthermore, they should be identical to each other and to the original double-stranded molecule. Be sure that you understand the process of replication before you go on to the next part of the exercise—transcription.

Transcription: Synthesis of RNA

Transcription is the process of synthesizing RNA. Three forms of RNA are produced by transcription: messenger RNA (mRNA), transfer RNA (tRNA), and ribosomal RNA (rRNA). In this exercise, we examine the formation of mRNA only, but remember that transfer RNA and ribosomal RNA are formed in much the same way. During transcription, RNA nucleotides base-pair with DNA nucleotides on the gene side of the DNA molecule.

To prepare for the production of ribonucleic acids during the process of transcription, we must first synthesize RNA nucleotides. Remember that RNA nucleotides have ribose sugar in place of deoxyribose sugar and the uracil base in place of thymine.

1. Retain one of the two copies of double-stranded DNA you have just made. Break the other DNA molecule down into its smallest subunits (phosphate, sugar, base). Some of these subunits are now needed to assemble RNA nucleotides. (DNA is not disassembled to make RNA in cells, but we need to do this with this model so that we have enough pieces to make the RNA we need.)

2. Construct RNA nucleotides using ribose, phosphate, and the bases adenine, guanine, cytosine, and uracil. You need

        5 separate adenine nucleotides,
        5 separate uracil nucleotides,
        4 separate guanine nucleotides, and
        4 separate cytosine nucleotides.

Be sure that all the phosphates are on the left side of the sugar (figure 34.5). These 18 nucleotides represent the RNA pool in the cell.

3. Separate the DNA double helix into two strands.

4. Use the RNA nucleotides in the pool and match the proper nucleotide from the RNA nucleotide pool with its partner on the gene side of the DNA only. Be sure that you use the

Figure 34.5

278

original DNA sequence as found in figure 34.3 as the gene. This side is the genetic code; the other side does not carry genetic information but is important in the replication process.

5. This newly constructed mRNA molecule should consist of nine nucleotides. The order of the RNA nucleotides in the mRNA you constructed is predetermined by the order of nucleotides along the coding strand (gene) of the DNA molecule.

6. Remove the mRNA molecule from the DNA strand and move the RNA to the side. Now put the two separated strands of DNA back together as they were. You have just simulated the process of transcription. The mRNA molecule that was formed has picked up the code from DNA. The DNA is intact and not damaged. The same DNA code can be used again for the transcription of additional mRNAs if necessary. In fact, it is common for more than one RNA transcript to be produced each time the cell expends energy to unwind and open up the DNA helix.

Translation: Synthesis of Proteins

During translation, the genetic message, which is coded by the order of bases in the DNA, is translated into the structure of a protein by determining the order of amino acids in the protein. Messenger RNA carries the message from the DNA in the nucleus to the ribosomes in the cytoplasm. Transfer RNA picks up amino acids in the cytoplasm and carries them to particular places on the mRNA called codons. A **codon** is a linear sequence of three bases on the mRNA. Ribosomal RNA is part of the structure of the ribosome. The mRNA and tRNA come together at the ribosome during the synthesis of proteins from individual amino acids. In an actual cell, the ribosomes are composed of proteins and rRNA. We will not manufacture a model of the ribosome but will use the box the model came in to simulate the ribosome as it moves down the mRNA.

For translation to occur in the cytoplasm of the cell, all three forms of RNA are needed. You will use the mRNA just constructed as well as three molecules of tRNA. Remember that tRNA molecules in the cell are formed in the same manner as the mRNA but at different gene sites on the DNA. You will make the tRNA molecules directly from the remaining plastic parts to save time. Also keep in mind that because the plastic pieces are not flexible, these tRNA models are different from the real thing. Our models lack a phosphate at one end and are much shorter than actual tRNA molecules. Actually a transfer RNA molecule is composed of a single strand of about 100 nucleotides that folds back on itself. One end of the tRNA has a coding portion called an anticodon at one end and binds to a specific amino acid at its other end. There are actually 64 possible kinds of tRNA, but we will use only three in this simple simulation.

Simulate the process of translation as follows:

1. Assemble the three tRNA models as they appear in figure 34.6.
2. Attach the appropriate amino acid to each of the three tRNA molecules (figure 34.7). Make sure that the H and OH units are attached to the amino acids.
3. Place the model's box on the first set of three nucleotides on the mRNA molecule (CUU) to simulate the presence of a ribosome. Pair the appropriate tRNA to the first three nucleotides (codon) on the mRNA. Since the three bases on the tRNA pair with the three bases (codon) on the mRNA, the bases on the tRNA are often referred to as the **anticodon**.

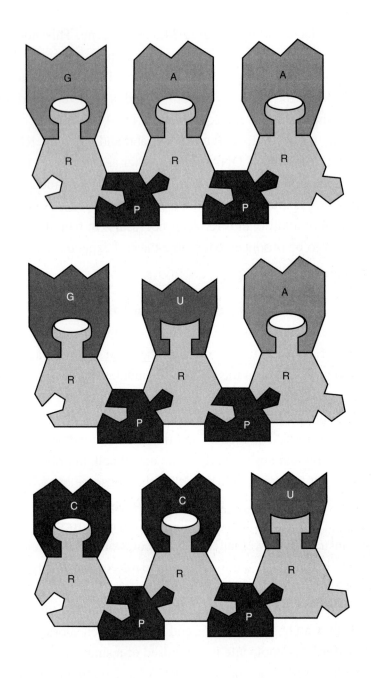

Figure 34.6

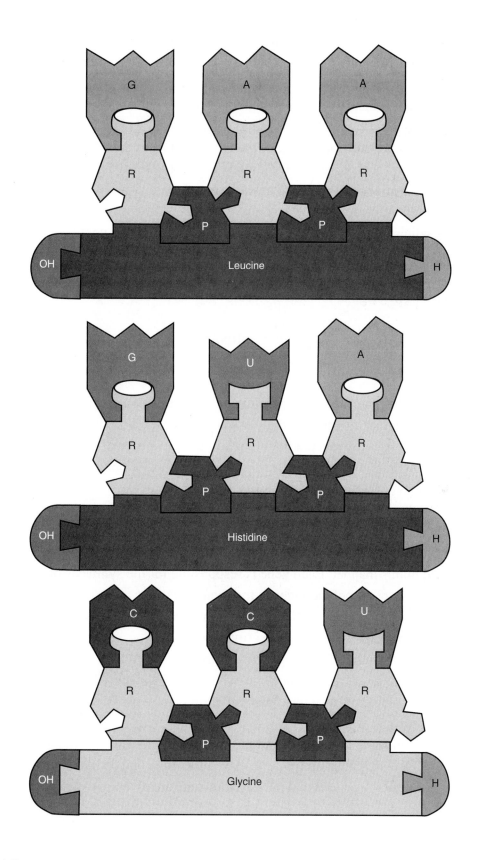

Figure 34.7

281

The process should occur as shown in figure 34.8.

4. Next, attach the appropriate tRNA to the second mRNA codon by matching the mRNA codon to its complementary tRNA anticodon. Note that the two amino acids are brought together so that the H end of one overlaps the OH of the other.

5. The two amino acids that have been brought adjacent to one another become chemically bonded to each other. Bonds between amino acids are called peptide bonds. To simulate the formation of a peptide bond, remove the H and OH from the adjacent amino acids and bond the two amino acids together. Link the H and OH to form a molecule of water. Because a water molecule is removed as the two amino acids are joined, it is often called a dehydration synthesis reaction.

6. Once this bonding has occurred, the ribosome (box) shifts to the next codon on the mRNA, and the anticodon of the next tRNA can be aligned with the codon on the mRNA. The first tRNA (GAA) is removed from the site and can bond with another leucine present in the cell cytoplasm. The original leucine molecule stays as part of the growing polypeptide chain.

7. When the third tRNA (CUU) binds to its mRNA codon, a second dehydration synthesis takes place, forming another peptide bond.

The completion of this process should result in the synthesis of a short polypeptide composed of the three amino acids: leucine, histidine, and glycine, in that order. You should also be able to identify three separated tRNA molecules ready to function in the transfer of more amino acids, two water molecules from dehydration synthesis reactions, and two peptide bonds holding the amino acids together.

Transfer RNA molecules may be used over and over again as amino acid–carrying molecules. A messenger RNA molecule can be used several times to produce the same polypeptide, but then it is broken down into its parts, which also return to the pools. All polypeptides, no matter how long they may be, are constructed in this manner. Each gene is responsible for the synthesis of a particular polypeptide that differs from other polypeptides in the sequence of the amino acids it contains and the length of the sequence. Each piece of DNA that contains information for the building of a particular sequence of amino acids is called a **structural gene**.

Table 34.2 shows the standard flow of genetic information from the DNA gene sequence in the nucleus to mRNA, which travels to the ribosome, where mRNA meets tRNAs, which carry the amino acids. The specific complementary **base pairing** allows one to work forward or backward along this informational transfer scheme.

The genetic dictionary, a list of codons and the specific amino acids they code for, has been determined (table 34.3). Note that most amino acids have more than one codon and that, in this particular dictionary, both mRNA codons and tRNA anticodons are given.

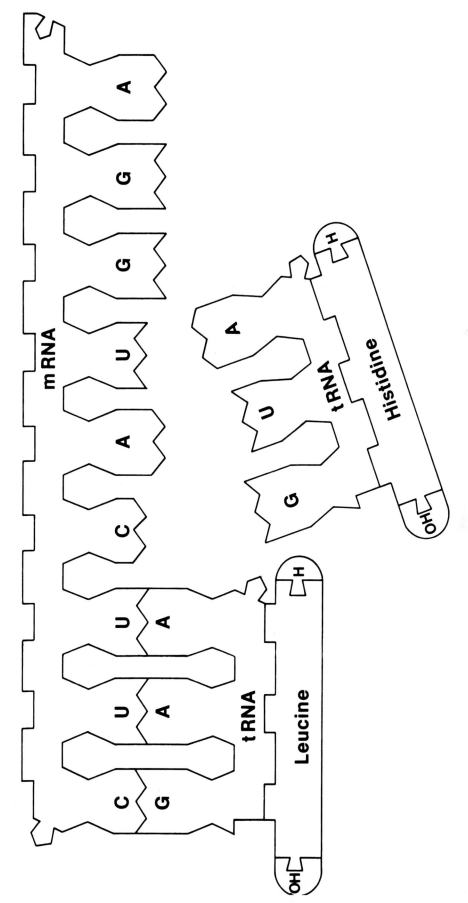

Figure 34.8

283

<table>
┌───┐
│ Table 34.2   Complementary Base Pairing   │
└───┘
</table>

DNA sequence		mRNA codon	tRNA anticodon	Carrying amino acid
C	G	C	G	
T	A	U	A	leucine
T	A	U	A	
C	G	C	G	
A	T	A	U	histidine
T	A	U	A	
G	C	G	C	
G	C	G	C	glycine
A	T	A	U	
complement	gene			

Results

Directions: Use the complete codon dictionary (table 34.3) to answer these questions.

1. Use the base sequence for mRNA to complete the columns on the following table (table 34.4). (Use the mRNA sequence shown in the table. Do not use the mRNA base pair sequence from the exercise you just completed.) Remember that complementary base pairing is the key. Refer to table 34.2 if you have problems.

Table 34.4 Complementary Base Paring			

DNA		mRNA	tRNA	Amino acid
___	___	U	___	
___	___	U	___	
___	___	U	___	___
___	___	A	___	
___	___	U	___	
___	___	C	___	___
___	___	U	___	
___	___	G	___	
___	___	U	___	___
complement	gene			

| Table 34.3 | Genetic Dictionary |

Amino acid	mRNA codons	tRNA anticodons	Amino acid	mRNA codons	tRNA anticodons
Phenylalanine	UUU	AAA	Tyrosine	UAU	AUA
	UUC	AAG		UAC	AUG
Leucine	UUA	AAU	Histidine	CAU	GAU
	UUG	AAC		CAC	GUG
	CUU	GAA	Glutamine	CAA	GUU
	CUC	GAG		CAG	GUC
	CUA	GAU	Asparagine	AAU	UUA
	CUG	GAC		AAC	UUG
Isoleucine	AUU	UAA	Lysine	AAA	UUU
	AUC	UAG		AAG	UUC
	AUA	UAU	Aspartic acid	GAU	CUA
Methionine	AUG	UAC		GAC	CUG
Valine	GUU	CAA	Glutamic acid	GAA	CUU
	GUC	CAG		GAG	CUC
	GUA	CAU	Cysteine	UGU	ACA
	GUG	CAC		UGC	ACG
Serine	UCU	AGA	Tryptophan	UGG	ACC
	UCC	AGG	Arginine	CGU	GCA
	UCA	AGU		CGC	GCG
	UCG	AGC		CGA	GCU
	AGU	UCA		CGG	GCC
	AGC	UCG		AGA	UCU
Proline	CCU	GGA		AGG	UCC
	CCC	GGG	Glycine	GGU	CCA
	CCA	GGU		GGC	CCG
	CCG	GGC		GGA	CCU
Threonine	ACU	UGA		GGG	CCC
	ACC	UGG	Terminator	UAA	
	ACA	UGU		UAG	
	ACG	UGC		UGA	
Alanine	GCU	CGA	Initiator	AUG	
	GCC	CGG			
	GCA	CGU			
	GCG	CGC			

2. Use one word to describe the relationship between the gene sequence and the mRNA sequence in table 34.4.

3. Describe in a sentence the relationship between the gene sequence and the tRNA sequence in table 34.4.

4. What will the sequence of mRNA nucleotides be if the following represents the bases in a DNA molecule of a structural gene?

DNA gene   A A T G G T C C A C C G C T G
           | | | | | | | | | | | | | | |

mRNA   ☐☐☐☐☐☐☐☐☐☐☐☐☐☐☐

5. If a structural gene contains 300 DNA nucleotides, how many amino acids will be used in the protein synthesis process?

6. If a protein has 150 amino acids, how many DNA nucleotides would make up the structural gene?

7. A protein has the following amino acid sequence. Construct a DNA nucleotide sequence of the structural gene.

Phenylalanine—Glycine—Glycine—Alanine—Proline—Valine—Asparagine—Alanine

8. Compare your DNA sequence from question 7 to that prepared by the others in the lab. Are there any variations? Did such variations occur in your answer to question 4? What is the reason for the difference? Are there any advantages to these variations? What are they?

9. Fill in table 34.5

10. Was the purpose of this lab accomplished? Why or why not? (Your answer to this question should show thoughtful analysis and careful, thorough thinking.)

Nucleic acid type	DNA	mRNA	tRNA
Type of sugar present			
Bases present			
Number of phosphates present			
Function			
Describe the shape of the molecule			
Where can this nucleic acid be found?			

Table 34.5　　　Nucleotide Components and Function

Experiment 35: Mitosis — Cell Division

Invitation to Inquiry

Many people who have been diagnosed with certain kinds of cancers undergo chemotherapy. There are four generally recognized types of chemotherapeutic drugs.

1. Antimetabolites
2. Topoisomerase inhibitors
3. Alkylating agents
4. Plant alkaloids

Do an internet search and discover how each of these chemotherapies interferes with the cell cycle to control dividing cancer cells.

Background

All large organisms are composed of many cells. Growth of many-celled organisms involves an increase in the number of cells followed by an increase in size of the new cells. This is the basic mechanism by which a body grows or wounds are repaired. Cell division involves two major events: the distribution of identical copies of the genetic information from the parent cell to two daughter cells followed by cytoplasmic division. As a result of cell division, all the cells of a multicellular organism have the same genetic information. Notice in figure 35.1 that the parent cell divides by mitosis, producing two daughter cells. These two identical daughter cells contain exactly the same genetic material as the parent cell. Mitosis assures the production of identical sets of genetic information in the daughter cells.

Mitosis is an orderly series of events that results in the equal distribution of the **chromosomes** that carry the genetic information to the two new cells. The process flows from one stage to the next without interruption. Traditionally, mitosis has been artificially divided into phases:

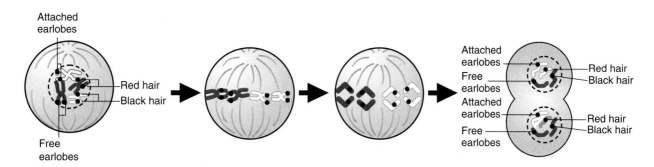

Figure 35.1

prophase, metaphase, anaphase, and **telophase.** It is a convenience to be able to classify parts of the process in this way, but it can be misleading if we forget that there is really no pause or interruption in the events. It might be helpful to consider these four phases as being pictures taken with a flash camera. In such pictures, the motion is frozen so that we can examine details that may interest us. Similarly, when we look at a plant or animal cell that has been killed and stained while in the process of mitosis, we are looking at a cell that was at a particular point in the mitotic process.

It is also important to recognize that mitosis is a relatively brief period in the life of a cell. A typical mitosis will take 2 to 4 hours to occur. The majority of the life of a cell is spent in a nondividing condition known as **interphase.** During interphase, the cell is participating in many important activities, including growth and the replication of DNA. Many cells differentiate into specialized cells that do not divide. They remain permanently in interphase, and they synthesize molecules, move, or perform other activities typical of the cell. Figure 35.2 shows the cell cycle and the part mitosis plays in it.

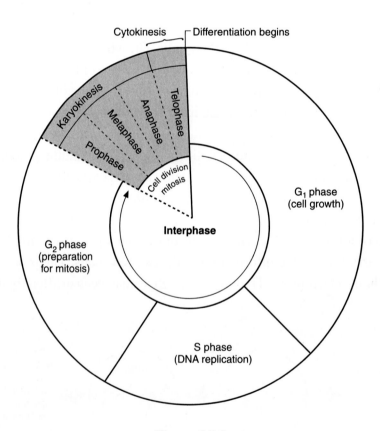

Figure 35.2

290

Procedure

1. Obtain a slide of *Allium* (onion) root tip or other plant material provided. Refer to figure 35.3 to orient yourself to the plant tissues you will study. The cells of the onion root tip were killed and stained. Then the root tip was sliced lengthwise into thin strips in which you will be able to see cells that were in different stages of the cell cycle when they were killed and stained.
2. Locate the tip of a root on low power and focus. Then switch to high power. It is important that you have clear focus using high power before you attempt to identify the various stages of mitosis.

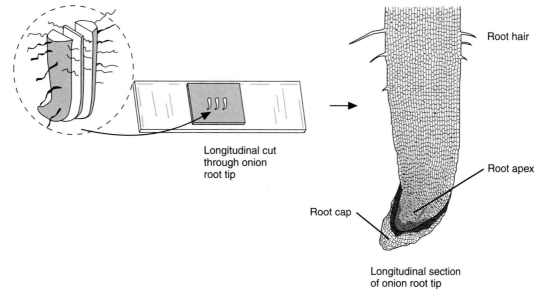

Root hair

Longitudinal cut
through onion
root tip

Root apex

Root cap

Longitudinal section
of onion root tip

Figure 35.3

3. You are now ready to scan your slide to locate the stages of mitosis in plant tissue. It is easiest to proceed in a step-by-step fashion, reading about the key events in the division of a "typical" cell in the paragraphs that follow before looking for that actual cell type. As you read, compare the description of the events with the photographs in figures 35.5 and 35.6.

Onion Root Tip

Interphase

Interphase is an active metabolic stage during which the cell performs its normal functions. It carries on metabolism, grows, and replicates DNA during this stage. However, an interphase cell is not dividing. The nucleus contains chromosomes, but they are in a tangled mass of threads, which presents a uniform appearance of tiny dots. Complete chromosomes cannot be seen. The nuclear membrane is present and one or two **nucleoli** are visible in the nucleus. During this stage, the DNA replicates, but these molecules are much too small to be visible. Replication is necessary if the cell is going to divide at some time in the future. Figure 35.4 shows the diagram of a chromosome before and after replication.

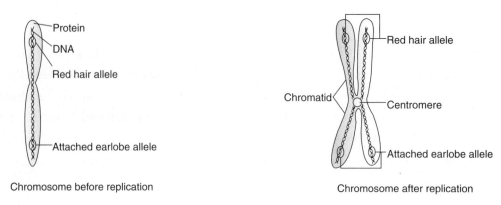

Chromosome before replication

Chromosome after replication

Figure 35.4

1. Locate a cell that you think is in the interphase stage and place the cell at the end of your pointer. Call the instructor over to check it. Let your neighbors look through your microscope while you examine the cell that they have selected as an example of interphase. When you are sure you can identify an interphase cell proceed to find a prophase cell.

Prophase

During prophase, the cell prepares itself for division. One of the preparations is the shortening and thickening of the chromosomes into structures that can be seen. Their first appearance is as a mass of fine, tangled threads. The chromosomes shorten and thicken as a result of coiling. They gradually become more easily seen as individuals. Although the chromosomes and DNA were replicated in interphase it is not possible to see that the chromosomes are double structures until late prophase. By late prophase you should be able to see that the chromosomes consist of two **chromatids** joined at a point called the **centromere**. Meanwhile, a series of tiny tubules forms a structure called the **spindle** (figure 35.5).

At the same time the nucleoli disappear. Eventually, at the very end of the prophase stage, the nuclear membrane disintegrates, and the chromosomes are lying free in the cytoplasm.

2. Look for a cell showing prophase. Again, place the cell at the end of the pointer in your microscope and share your slide with your neighbors. If you have any doubts, ask your instructor to check the slide.

Metaphase

During metaphase, the chromosomes become attached by their centromeres to spindle fibers. The chromosomes appear to be tugged by the spindle fibers so that they form a flat sheet of chromosomes on a plane that passes through the equator of the cell. Viewed from the side, they appear to form a line; viewed from the pole, they can be seen to form a ring or disk. (Onion slides will not show polar views because all the cells have a lengthwise orientation in the root tip.)

3. Look for a cell showing metaphase. Again, place the cell at the end of the pointer in your microscope and share your slide with your neighbors. If you have any doubts, ask your instructor to check the slide.

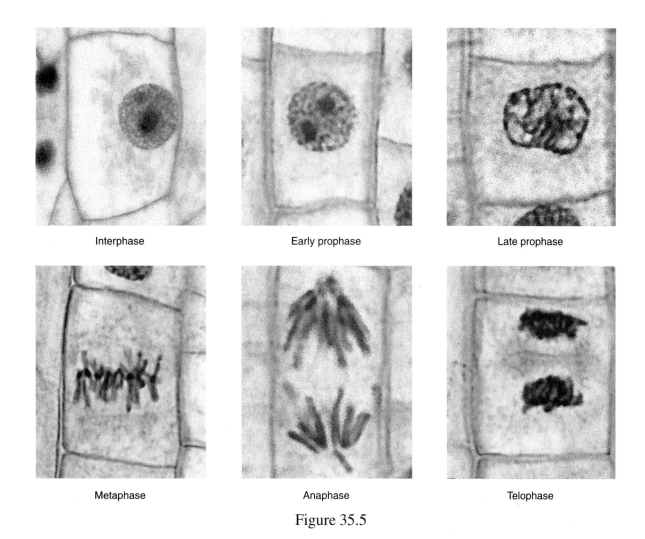

Interphase	Early prophase	Late prophase
Metaphase	Anaphase	Telophase

Figure 35.5

Anaphase

It is during anaphase that the equal distribution of genetic information is accomplished. Each centromere divides, and the two identical chromatids of each chromosome are pulled to opposite sides (poles) of the cell by the spindle fibers. When the two chromatids of a chromosome separate, each is called a daughter chromosome.

4. Look for a cell showing anaphase. Again, place the cell at the end of the pointer in your microscope and share your slide with your neighbors. If you have any doubts, ask your instructor to check the slide.

Telophase

A number of changes occur during the telophase stage of division. The chromosomes form two groups at opposite ends of the cell and the individual chromosomes become difficult to see as they uncoil. The spindle apparatus breaks down, and a nuclear membrane forms around each group of chromosomes. These are now called **daughter nuclei**. Nucleoli also reform within the nucleus. As cell division is completed, **cytokinesis** (division of the cytoplasm) occurs by the formation of a **cell plate** between the two daughter nuclei. If you find evidence of the plate being formed, you can be sure that the cell is in telophase.

5. Look for a cell showing telophase. Again, place the cell at the end of the pointer in your microscope and share your slide with your neighbors. If you have any doubts, ask your instructor to check the slide.

Whitefish Mitosis

Obtain slides of whitefish blastula cells showing mitosis. A blastula stage is an embryonic stage that is shaped like a ball of cells. Because embryos grow rapidly they must be made of cells that divide frequently. The blastula has been sliced into thin disks and stained to help you see the chromosomes and other structures. The slide will have several disks on it, and you will probably need to look at all of them to see the various stages of the cell cycle. Locate a section of a whitefish blastula under low power and focus; then switch to high power. Refer to figure 35.6.

It is often difficult to clearly distinguish interphase cells in the whitefish blastula slide preparation. This is because the cells divide so rapidly that interphase is a very short period of time; thus few cells are in interphase. Sections like the ones you are viewing can be sliced at any angle through a given cell. Therefore, some of the cells will be sliced so that you will see the cell from the equator, whereas others will show the cell from the pole. You may also find cells that do not appear to have any chromosomes because the knife removed all the chromosomes and left an empty cell. Or the slice of cell may have included only part of the chromosomes. If you look for the spindle fibers, you may be able to figure out the orientation more readily.

1. Locate a prophase stage. Share it with others. Check with your instructor if you have trouble.

2. Locate an equatorial view of metaphase. You should be able to see the spindle fibers radiating from the poles and the chromosomes lined up at the equatorial plane.

3. Locate a polar view of the metaphase. Note that this view does not show either of the poles or any of the spindle fibers. The chromosomes are grouped in a ring or disk but are not enclosed by a nuclear membrane. Figure 35.6 is especially helpful in identifying this view.

4. Locate a cell in the anaphase stage.

5. Locate a telophase cell.

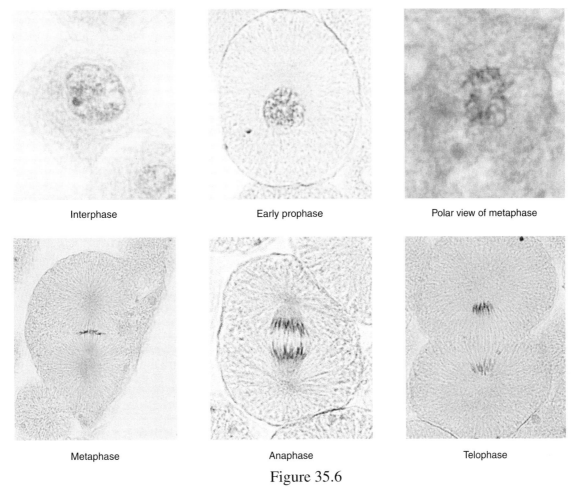

Interphase   Early prophase   Polar view of metaphase

Metaphase   Anaphase   Telophase

Figure 35.6

Cooperative Trial Quiz

Choose one stage of either plant or animal mitosis and put your pointer on it. Pick a cell that you think is a good example of a particular stage of mitosis. Write the name of this stage on a slip of paper. On the opposite side of the slip of paper, write the number you are assigned by your instructor. Put the slip by your scope so that the number side is up and the side with the name of the stage is down. Use the Mitosis Quiz Answer Sheet found at the end of this lab. Visit each student's scope and identify the stage at the end of the pointer. Your instructor will also take the trial quiz and make a key. If you have made any errors, be sure to study the cells you identified incorrectly so that you do not make the same mistake again.

DNA-Mitosis Relationship

The chromosomes found in cells are made up of DNA and protein molecules. Both chromatids of a chromosome contain identical DNA molecules. Different chromosomes have different sequences of DNA. Therefore you should be able to follow DNA molecules as you proceed through the cell cycle. The first circle on the next page represents the outline of the cell that was just formed from mitosis and is entering interphase. It contains four single-stranded chromosomes with a sketch of DNA superimposed on them. In the circles that follow, sketch a series of views of this cell as it proceeds through interphase and enters mitosis (prophase, metaphase, anaphase, and telophase). Show what happens to the chromosomes and the DNA.

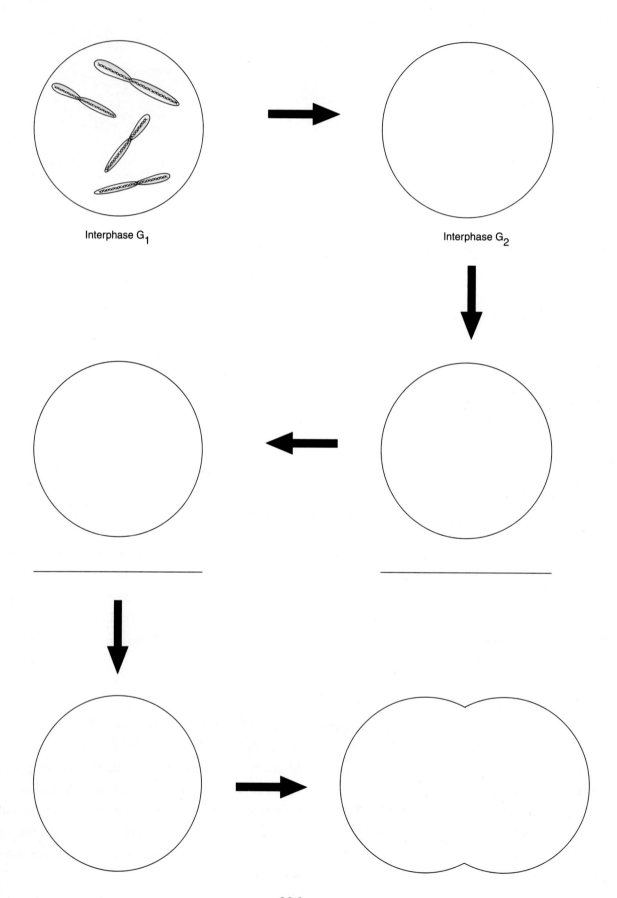

Interphase G₁

Interphase G₂

Results

1. During which part of the cell cycle does DNA replication occur?

2. When do chromosomes first become visible in mitosis?

3. What is the difference between the terms chromosome and chromatid?

4. What is a spindle? What is its function?

5. During what stages of mitosis are chromosomes composed of two chromatids?

6. During what stages of mitosis are chromosomes single structures composed of one chromatid?

7. How does cytokinesis differ in plant and animal cells?

8. Most cells spend the longest amount of time in _____. What evidence do you have to support this statement?

9. Why was it difficult to find interphase cells in the whitefish blastula slide?

10. Why are you more likely to see a polar view in animal cells than in plant cells?

11. Was the purpose of this lab accomplished? Why or why not? (Your answer to this question should show thoughtful analysis and careful, thorough thinking.)

Mitosis Quiz Answer Sheet

1._____

2._____

3._____

4._____

5._____

6._____

7._____

8._____

9._____

10._____

11._____

12._____

13._____

14._____

15._____

16._____

17._____

18._____

19._____

20._____

21._____

22._____

23._____

24._____

25._____

Experiment 36: Meiosis

Invitation to Inquiry

In some boy babies the testes are retained within the abdomen and do not descend into the scrotum. The temperature of the scrotum is lower than that of the abdomen. If this situation is not corrected surgically, the person will be sterile because normal spermatogenesis only occurs at the lower temperature present in the scrotum.

Some cases of male sterility have been linked to a man's wish to wear tight pants that kept the testes close to his body and thus elevated their temperature. With this information in mind, what other normal activities or events might have the same effect?

Background

Nearly all organisms have life cycles in which sexual reproduction occurs. Sexual reproduction involves the joining of sex cells (**gametes**) from two different parent organisms to produce a new individual. The joining of sex cells is called **fertilization**, and the new cell formed from their union is called a **zygote**. Because two gametes join to form one zygote during fertilization, the zygote must have twice as many chromosomes as either of the gametes that formed it. The zygote is said to have the **diploid** (two sets) number of chromosomes and the gametes the **haploid** (one set) number of chromosomes. Often the diploid cells are designated as $2n$ because they have two sets of chromosomes; the haploid cells are designated as $1n$ or n because they have only one set of chromosomes. If a sexually reproducing species is to retain a fixed number of chromosomes, the parents must have a mechanism to produce gametes with half the number of chromosomes typical for the organism (see figure 36.1).

Meiosis is a special kind of cell division in which the cells produced (gametes) have half the number of chromosomes typical for the species. Meiosis occurs only in the gonads (testes and ovaries) of sexually reproducing organisms. The process of meiosis involves two divisions with the first division (meiosis I) immediately followed by the second (meiosis II). During the first division of meiosis (meiosis I), the chromosomes join in homologous pairs. **Homologous chromosomes** are the same length and carry genetic information for the same characteristics. One chromosome of each pair originated from each of the parents of the individual in which meiosis is occurring. The pairing of homologous chromosomes is known as **synapsis**. Following synapsis, the pairs of homologous chromosomes line up at the equator of the cell. They then separate, and one member of each pair moves to each pole of the cell. When the two nuclei reorganize and cytokinesis occurs, the resultant two cells have one half as many chromosomes as the original cell. Reduction from the diploid number ($2n$) of chromosomes to the haploid number (n) of chromosomes has been accomplished.

Each of the two cells produced as a result of meiosis I goes through a second division. This second division (meiosis II) results in the formation of four cells. During this division, individual

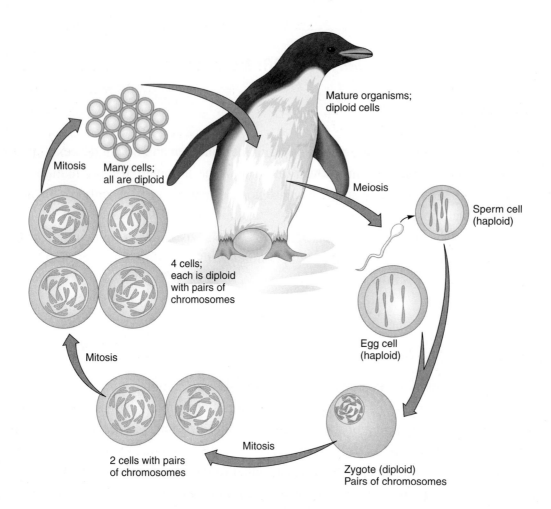

Figure 36.1 The cells of this adult penguin have, for our purpose, eight chromosomes in their nuclei. In preparation for sexual reproduction, the number of chromosomes must be reduced by half so that fertilization will result in the original number of eight chromosomes in the new individual. The offspring will grow and produce new cells by mitosis.

chromosomes line up on the equators of the cells. The centromeres divide, and one of each of the chromatids (daughter chromosomes) moves to each pole of the cell.

In males, the four cells that result from these meiotic divisions mature into sperm cells. Each cell has the haploid number of chromosomes, and each daughter chromosome (chromatid) is a single structure that carries a single DNA molecule. Each cell contains a complete set of genetic information for all the characteristics of an organism, but the combination of characteristics in each gamete is different from that in others. In females only one of the four cells matures into an egg. The other three cells die. However, the same processes that generate variety in the kinds of sperm produced also are involved in producing eggs so that eggs are as different from one another as are sperm.

There are several other terms that need to be clarified so that you can more easily follow the exercise. A **gene** is a piece of DNA that directs the expression of a particular characteristic. An **allele** is a gene for which there is an alternative expression. We generally use the word *gene* when we are talking about characteristics in general and use the word *allele* when we are talking about specific examples of genes for which there are several alternatives. For example we can talk about hair color genes, but when we want to talk specifically about the different possible hair colors we talk about the several alleles for hair color: red, blond, brown, and black. Some alleles are **dominant** to others and mask or cover up the presence of other alleles. The alleles that are masked are called **recessive** alleles. Each diploid organism has two alleles for each characteristic; one was received from each parent. The alleles may be identical (two alleles for red hair) or they may be different (one allele for red and one allele for black). When the two alleles are identical, the organism is said to be **homozygous**. When the two alleles are different, the organism is said to be **heterozygous**. The **genotype** of an organism is a *listing* of the two alleles for each trait that it possesses. The **phenotype** of an organism is a *description* of the way a characteristic is displayed in the structure, behavior, or physiology of the organism.

During this lab exercise you will

1. determine the phenotype of a parent from the alleles on models of chromosomes.
2. manipulate the chromosome models to simulate synapsis.
3. follow the chromatids as they cross over.
4. simulate metaphase and anaphase stages of meiosis I.
5. manipulate the chromosome models as they proceed through meiosis II and produce four gametes.
6. unite one of your four gametes with one of the four gametes from another group of students to form a zygote.
7. compare the phenotype of your zygote with the phenotypes of other zygotes and with the parental phenotype.

Procedure

To understand the mechanism involved in the production of gametes, we will manipulate models of chromosomes. Each chromosome is composed of two identical chromatids.

1. Work in pairs. Obtain four model chromosomes that have already completed DNA replication and are, therefore, composed of two identical chromatids. These four chromosomes should include a short pair of chromosomes with two alleles and terminal centromeres and a longer pair of chromosomes with four alleles. The centromere of this long pair of chromosomes is more centrally located. The two chromosomes of each pair should be different colors.
2. Place the four chromosomes randomly on a large sheet of paper to represent their presence in the cytoplasm of a cell.
3. Note the genes on the short pair of chromosomes. At one point (**locus**) on each chromosome there is information about insulin production. At a different locus is information about hair color. Because these chromosomes are the same size, have their centromeres in the same relative position, and have genes for the same characteristic at equivalent points along their lengths, they

are called **homologous chromosomes**. The various alternative expressions of genes on homologous chromosomes (dark hair or light hair, insulin production or no insulin production) are known as **alleles**.

4. Compare your four chromosomes with the diagrams in figure 36.2 to make sure that you start with a proper set of information. Make sure that you have the correct alleles on the correct chromosomes and that you have the correct number of beads present between alleles. If your chromosomes are not as shown in figure 36.2, make corrections before proceeding. Notice that on the figure and on your chromosome models, some of the genetic information is in capital letters and some is in lowercase letters. Those alleles that are **dominant** (those that always express themselves) are in capital letters. **Recessive** alleles (those expressed only when no dominant allele is present) are in lowercase letters.

5. Let us suppose that the models of the two pairs of homologous chromosome represent the complete set of genetic information of an organism. (In reality humans have 46 chromosomes and thousands of genes.) This cell was formed by the uniting of a sperm and an egg. One set of chromosomes is from the father and one set from the mother. Let the chromosomes with the darker beads represent the chromosomes donated by the father and those with the lighter beads represent the chromosomes donated by the mother. The cell you are dealing with has two sets of genetic information (2n) with one allele for each characteristic from the father and the other from the mother. In table 36.1, list the genotype and phenotype of this organism. Remember the genotype is a listing of the two alleles for each characteristic and the phenotype is a description of the characteristics displayed in the structure, behavior, or physiology of the organism.

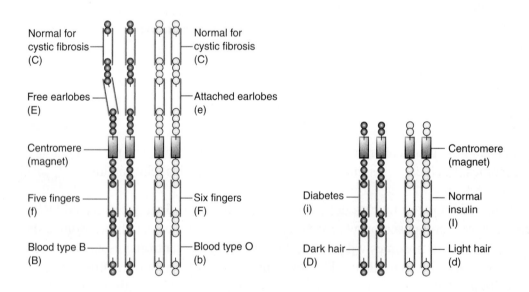

Figure 36.2

Table 36.1	List of Characteristics		
Trait	Genotype of Organism		Phenotype of Organism
	From Mom	From Dad	
Cystic fibrosis	C	C	Normal for CF
Ear shape			
Finger number			
Blood type			
Insulin production ability			
Hair color			

Prophase I (Synapsis and Crossing-Over)

Because meiosis consists of two divisions, the various stages of the process are labeled I or II to indicate whether they are occurring in meiosis I or meiosis II. During the prophase I stage of meiosis, the nucleus prepares for a division. The chromosomes become visible, the nuclear membrane begins to disintegrate, the nucleolus disappears, and the spindle apparatus develops. In addition, during prophase I homologous chromosomes pair with each other along their entire length. While they are paired like this, we say they are in **synapsis**.

1. Put the two members of each homologous pair of chromosomes near each other to simulate their synapsis. While in the synapsed condition, equivalent pieces of homologous chromatids will be broken off and exchanged. This process, called **crossing-over**, can occur several times along the length of homologous chromosomes.
2. Simulate crossing-over in your models by detaching, exchanging, and reattaching exactly equal parts of chromatids between the two members of a homologous pair of chromosomes. (The pop beads will allow you to break a chromatid in a variety of places, but for each crossover you will need to make the break at the same place on equivalent chromatids.) Each chromatid acts as an independent unit in this process. Therefore, you should not make exactly the same changes to both chromatids of a chromosome. Make a minimum of two crossovers for each pair of homologous chromosomes. You will now have chromosomes that contain alleles that were originally on the other member of the homologous pair, and each chromosome will contain both colors of beads. These are the chromosomes you will use for the remainder of the exercise.
3. As a result of the process of crossing-over, a new combination of alleles has been formed. Look at figure 36.2 and note the differences between the arrangement of alleles in the original chromosomes and the chromosomes you have just crossed over.
4. Compare your chromosome models with the models of other students. Are their models like yours? Do they need to be?

If you do not understand the process of crossing-over, ask your instructor for help before going on.

Metaphase I (Alignment of Chromosome Pairs)

The chromosome pairs, still in synapsis, line up at the equatorial plane.

1. Move your model chromosomes on your piece of paper to show this arrangement. When the chromosomes are in this position, the cell is in metaphase I.
2. How does the arrangement of chromosomes in metaphase I of meiosis differ from metaphase in mitosis? In the space below, sketch the arrangement of four chromosomes in metaphase I of meiosis and the same four chromosomes in metaphase of mitosis.

Anaphase I (Independent Assortment and Segregation)

During this stage, the separation of homologous chromosomes occurs. This separation is called segregation. Segregation is the separating of homologous chromosomes to opposite poles of the cell. Because the chromosomes carry genes, the genes are also separated (segregated) into two sets during anaphase I.

1. Segregate the two members of one pair of homologous chromosomes by moving their centromeres to opposite ends of the paper. Similarly segregate the other pair. Segregation of one pair of chromosomes is independent of the segregation of the other pair. The way a pair of homologous chromosomes happens to be aligned at the equator determines the direction in which each moves. The fact that the alignment and segregation of one pair of chromosomes is independent of the alignment and segregation of other pairs of chromosomes is called **independent assortment**. Because each set of homologous chromosomes is carrying genes, the genes on nonhomologous chromosomes are segregated independently of one another.
2. How do the chromosomes at the end of anaphase I of meiosis differ from the chromosomes at the end of the anaphase stage of mitosis? In the space below, sketch the arrangement of four chromosomes in anaphase I of meiosis and the same four chromosomes in anaphase of mitosis.

Telophase I (Cytokinesis)
During this stage, cytokinesis occurs and results in the formation of two daughter cells.

1. Show this division by tearing your paper into two equal parts. Each part will represent half the original cytoplasm. Each of these daughter cells has one chromosome from each of the two homologous pairs and is haploid. Each has half as many chromosomes as the original cell, but both have a complete set of genetic information. Separate the two sets of chromosomes by enough distance to remind you that they are in different cells.
2. Compare the sets of genetic data in each of the two cells by listing the alleles present on the chromosomes in table 36.2.

Table 36.2   Comparison of Genetic Data of Cells Resulting from Meiosis I

Genetic Data, Daughter Cell-1	Genetic Data, Daughter Cell-2
Longer chromosome Shorter chromosome	Longer chromosome Shorter chromosome

3. Compare your daughter cells to those formed by other students. Do both have the same combination of alleles in each cell? _____

4. Is it necessary that they be the same? _____

5. List two meiotic processes that contributed to these differences.

Prophase II (Cells Prepare for Second Division)
No replication of chromosomes or DNA occurs after telophase I, and each of the two cells enters prophase II. The activities that occur in prophase II are the same as those that happen in prophase of mitosis. The chromosomes become visible, spindle fibers form, and the nuclear membrane breaks down.

1. List three ways that a prophase II cell differs from a prophase I cell.

Metaphase II (Alignment of Chromosomes at Equator)

During metaphase II, the chromosomes in each cell are lined up on the equatorial plane. They are not in homologous pairs; each chromosome, however, is still made up of two chromatids.

1. Move your chromosome models to the center of the half sheet of paper to represent this activity in both daughter cells.
2. The original diploid number of chromosomes in your cell was four. How many chromosomes can be found in a daughter cell in metaphase II? _____

Anaphase II (Chromatids Separate)

During anaphase II, the centromeres split, and the chromatids (now referred to as daughter chromosomes) migrate toward opposite poles.

1. Separate your chromatids (daughter chromosomes) at the centromere and move them toward opposite poles (opposite ends of the paper). Do this in each of the two daughter cells.
2. List two ways in which anaphase I differs from anaphase II.

3. How does anaphase II of meiosis compare with anaphase of mitosis?

Telophase II (Haploid Gametes Formed)

During telophase II, cytokinesis occurs in each cell. This results in the formation of four haploid cells called gametes.

1. Represent cytokinesis in each of the two daughter cells. Simulate this by tearing the paper in half.
2. In addition to cytokinesis, what other processes would you expect to occur to the chromosomes and nucleus in a telophase cell?

3. List the alleles now found in each of the four cells in table 36.3.

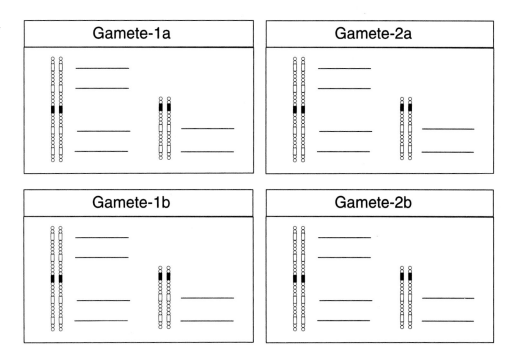

Table 36.3

4. Note that the four cells you have formed are different from one another. A number of things have happened during meiosis to cause these four cells to be different. Look back over meiosis I and II and identify the several processes that contribute to the differences in these four cells. List them in the space provided.

Fertilization (Joining of Gametes to Form a Zygote)

Your instructor will have designated each group in the class as being either a male or female.

1. A male group randomly selects one of its four sperm cells and delivers it to a female. The female group has previously selected at random one of its egg cells.
2. Unite these two gametes to represent **fertilization**. This fertilized egg cell is known as a **zygote**.
3. Record the genotype and phenotype of the offspring resulting from fertilization on table 36.4. Show the alleles it received from each parent and the phenotype that would be observable. Share your data with the other groups in class.

Table 36.4 Genotypic and Phenotypic Characteristics of Offspring

Trait	Genotype of Organism		Phenotype of Organism
	Allele from Mom	Allele from Dad	
Cystic fibrosis	C	C	Normal for CF
Ear shape	___	___	
Finger number	___	___	
Blood type	___	___	
Insulin production ability	___	___	
Hair color	___	___	

4. Compare the phenotype of your offspring with the phenotype of both parents.
5. Compare the phenotypes of all offspring produced in class.
6. Fertilization (the joining of two haploid gametes) results in a diploid zygote which will develop into a new individual organism. What effect does being diploid rather than haploid have on determining what the phenotype will be?

At the end of this exercise, arrange the chromosome models as they appear in figure 36.2. Have your instructor check the chromosomes before you leave the lab.

End-of-Exercise Questions

1. How many of the hypothetical offspring produced during this lab activity had the same phenotype?_____

   The same genotype? _____

2. How do the results of meiosis and mitosis differ in terms of chromosome numbers? Fill in the diagrams by assuming that each original cell represents a human cell with a diploid number of 46 chromosomes.

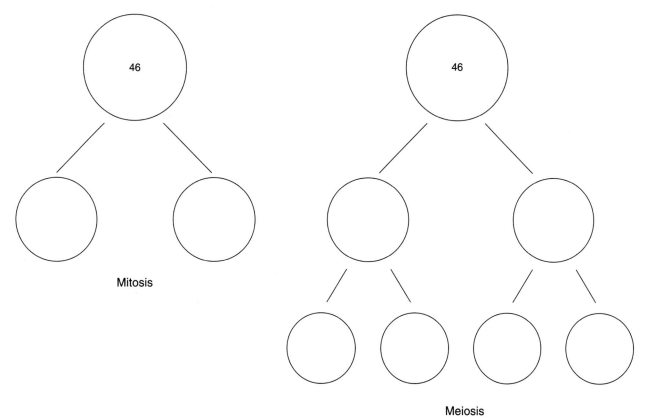

Mitosis

Meiosis

3. Many people use the terms *gene* and *allele* interchangeably. Are they the same? Explain.

4. What would you expect to be happening in any cell undergoing metaphase?

5. List the activities that occur during meiosis that contribute to the variety seen in the gametes produced.

6. Why is meiosis necessary in sexually reproducing organisms?

7. Name the event of meiosis that ensures that the chromosome number is reduced.

8. During which stage of meiosis does each of the following events occur?

    a. Synapsis _____

    b. Crossing-over _____

    c. Segregation of homologous chromosomes _____

    d. Reduction from diploid to haploid _____

    e. Separation of chromatids _____

    f. Independent assortment of homologous chromosomes _____

9. Many people have the misconception that a dominant allele must be the most common in a population. Address this misconception by using the finger number trait as an example.

Experiment 37: Genetics Problems

Invitation to Inquiry

Construct a family tree that involves at least 4 generations. Follow a particular characteristic through those 4 generations. If you have a particular genetic quirk in you family you may follow that and try to determine how it is inherited.

If you don't have any unusual characteristics in your family, earlobe shape is an easy characteristic to follow. Some people have an earlobe that hangs free while others have the lobe attached directly to the side of the face. You can even use photographs of your ancestors to determine earlobe shape. Ask yourself the following kinds of questions.

Does it show up in every generation? Is it present in both sexes? Is it a dominant or recessive characteristic?

Background

One of the curiosities of life is the variety of offspring that can be produced as a result of sexual reproduction. The types of traits possible in an offspring have long been of interest to humans. Some people are interested for personal reasons—a new baby due in the family; some people are interested for business reasons—the desire to breed a new variety of plant. In either case they want to know the probability of having a given type of offspring. Although offspring receive half their genes from each parent, because of chance combinations of genes, they may resemble one parent more than the other or may show characteristics that are not displayed by either parent.

To understand how characteristics are passed from one generation to the next we need to know some basic information. Every individual produced by sexual reproduction has two genes for each characteristic. They receive one from each parent. However, there are alternative genes for the same characteristic known as **alleles**. For example there are alternative genes for eye color; the blue eye allele and the brown eye allele. Some alleles called **dominant alleles** are able to mask the presence of other **recessive** alleles. If an individual has two identical alleles for a characteristic (two blue eye alleles or two brown eye alleles) it is **homozygous**. If the two alleles are different from one another (one brown eye allele and one blue eye allele) the individual is **heterozygous**. Therefore, an individual may have some recessive alleles that do not express themselves but are still part of their genetic catalog. All the genes that an individual has is its **genotype**. The observable characteristics displayed in the organism's structure, behavior, or physiology are known as the organism's **phenotype**.

To determine how characteristics are passed from one generation to the next we must first know something about the parents. What do they look like? What alleles do they possess? We must know what alleles are possible and the probability of each allele appearing in the gametes produced

311

by the parents. We must also know the possible ways these may combine during fertilization. In this exercise, you are asked to determine how genes are passed from one generation to the next and determine the genotypes and phenotypes of parents and offspring. During this lab exercise you will

1. work a probability problem.
2. work single-factor inheritance problems.
3. work double-factor inheritance problems.
4. determine genotypes of parents and offspring.

Probability Versus Possibility

To solve heredity problems, you must have an understanding of probability. Probability is the chance that a particular desired outcome will happen divided by the total number of possible outcomes. Probability is often expressed as a percent or a fraction. Probability is not the same as possibility. It is possible to toss a coin and have it come up heads. But the probability of getting a head is more precise than just saying it is possible to get heads. The probability of getting heads is one out of two (1/2 or 0.5 or 50%) because a coin has two sides, only one of which is a head. The desired outcome (heads) is divided by the total number of outcomes (heads or tails). This probability can be expressed as a fraction:

$$\text{Probability} = \frac{\text{the number of events that can produce a given outcome}}{\text{the total number of possible outcomes}} .$$

What is the probability of cutting a deck of cards and getting the ace of hearts? The number of times that the ace of hearts can occur is 1. The total number of possible outcomes, the number of cards in the deck, is 52. Therefore, the probability of cutting an ace of hearts is 1/52.

What is the probability of cutting an ace? The total number of aces in the deck is 4, and the total number of cards is 52. Therefore, the probability of cutting an ace is 4/52 or 1/13.

It is also possible to determine the probability of two independent events occurring together. The probability of two or more events occurring simultaneously is the product of their individual probabilities. For example, if you throw a pair of dice, it is possible that both will be a 4. What is the probability that both will be a 4? The probability of one die being a 4 is 1/6. The probability of the other die being a 4 is also 1/6. Therefore, the probability of throwing two fours is

$$1/6 \times 1/6 = 1/36.$$

| 1/6 | x | 1/6 | = | 1/36 chance for two successive fours |

Probability Problem

Probability is a mathematical statement about how likely something will occur. It is not certainty. To help you understand this concept we will work with a deck of playing cards and determine the likelihood of getting each of the four suits.

Work in pairs. One student shuffles and cuts a standard deck of cards. The other student records the suit. Repeat this 100 times; shuffling, cutting, and recording. Record your information in table 37.1.

Table 37.1   Probability Problem Results

Suit	Actual	Expected	Difference
Hearts		25	
Clubs		25	
Diamonds		25	
Spades		25	

Your instructor will tell you how to record this information for the whole class. Record the actual number of times each suit is cut. Compare your results with the entire class. Why is there a difference between the results you got and the results of the whole class?

Single-Factor Inheritance Problems

Single-factor crosses are concerned with how a single genetic trait is passed from the parents to an offspring. Solving a heredity problem requires five basic steps.

Step 1: Assign a symbol for each allele.
Usually, a capital letter is used for a dominant allele and a small letter is used for a recessive allele.

For example, use the symbol *E* for free earlobes, which is dominant, and *e* for attached earlobes, which is recessive.

$$E = \text{free earlobes}$$
$$e = \text{attached earlobes}$$

Step 2: Determine the genotype of each parent and indicate a mating.
Suppose both parents are heterozygous; the male genotype is *Ee,* and the female genotype is also *Ee.* The × between them is used to indicate a mating.

$$Ee \times Ee$$

Step 3: Determine all the possible kinds of gametes each parent can produce.

Remember that gametes are haploid; therefore, they can have only one allele instead of the two present in the diploid cell. Because the male has both the free earlobe allele and the attached earlobe allele, half his gametes contain the free earlobe allele, and the other half contain the attached earlobe allele. Because the female has the same genotype, her gametes are the same as his.

For solving genetics problems, a *Punnett square* is used. A **Punnett square** is a box figure that allows you to determine the probability of obtaining each of the genotypes and phenotypes possible in the offspring resulting from a particular cross. Remember, because of the process of meiosis each gamete receives only one allele for each characteristic listed. Therefore the male gives either an *E* or *e*; the female also gives either an *E* or *e*. The possible gametes produced by the male parent are listed on the left side of the square; the female gametes are listed on the top. In our example, the Punnett square would show a single dominant allele and a single recessive allele from the male on the left side. The alleles from the female would appear on the top:

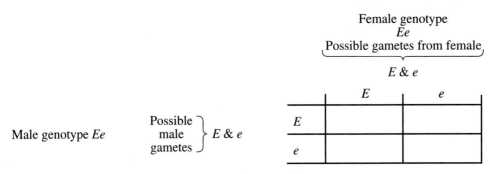

Step 4: Determine all the allele combinations that can result from the combining of gametes.

To determine the possible combinations of alleles that could occur as a result of this mating, simply fill in each of the empty squares with the alleles that can be donated from each parent. Determine all the allele combinations that can result when these gametes unite:

	E	e
E	EE	Ee
e	Ee	ee

Step 5: Determine the phenotype of each possible allele combination shown in the offspring.

In this instance, three of the offspring (those with the genotypes *EE*, *Ee*, and *Ee*) have free earlobes, because the free-earlobe allele is dominant to the attached-earlobe allele. One offspring, *ee*, has attached earlobes. Therefore, the probability of having offspring with free earlobes is 3/4 and with attached earlobes is 1/4.

Additional Single-Factor Inheritance Problems (One Trait Followed from One Generation to the Next)

1. In humans, six fingers (F) is the **dominant** trait; five fingers (f) is the **recessive** trait. Assume both parents are **heterozygous** for six fingers.
   What are the phenotypes of the father and the mother?
   What is the genotype of each parent?
   What are the different gametes each parent can produce?
   What is the probability of them having six-fingered children? five-fingered children?

   a. Father's phenotype_____; mother's phenotype_____

   b. Father's genotype_____; mother's genotype _____

   c. Father's gamete_____ or _____; mother's gametes_____ or _____

   d. Probability of a six-fingered child _____

   e. Probability of a five-fingered child _____

2. If the father is heterozygous for six fingers and the mother has five fingers, what is the probability of their offspring having each of the following phenotypes?

   Six fingers _____; five fingers _____

3. In certain flowers, color is inherited by alleles that show **lack of dominance (incomplete dominance)**. In such flowers, a cross between a **homozygous** red flower and a homozygous white flower always results in a pink flower. A cross is made between two pink flowers. Use F^w to represent the white allele and F^R to represent the red allele. What is the probability of each of the colors (red, pink, and white) appearing in the offspring?

4. Use the information given in the previous problem. A cross is made between a red flower and a pink flower. What is the expected probability for the various colors?

Double-Factor Inheritance Problems

A double-factor cross is a genetic study in which two pairs of alleles are followed from the parental generation to the offspring. These problems are basically worked the same as a single-factor cross. The main differences are that in a double-factor cross you work with two different characteristics from each parent.

It is necessary to recognize that **independent assortment** occurs when two or more sets of alleles are involved. Mendel's law of independent assortment states that members of one allelic pair separate from each other independently of the members of other pairs of alleles. This happens during meiosis when the chromosomes which carry the alleles segregate. (Mendel's law of independent assortment applies only if the two pairs of alleles are located on different pairs of homologous chromosomes. This is an assumption we will use in double-factor crosses.)

In humans, the allele for free earlobes dominates the allele for attached earlobes. The allele for dark hair dominates the allele for light hair. If both parents are heterozygous for earlobe shape and hair color, what genotypes and phenotypes can their offspring have, and what is the probability of each genotype and phenotype?

Step 1: Use the symbol *E* for free earlobes and *e* for attached earlobes. Use the symbol *D* for dark hair and *d* for light hair.

$$E = \text{free earlobes}$$
$$e = \text{attached earlobes}$$
$$D = \text{dark hair}$$
$$d = \text{light hair}$$

Step 2: Determine the genotype for each parent and show a mating.
In this example, the male genotype is *EeDd*, the female genotype is *EeDd*, and the × between them indicates a mating.

$$EeDd \times EeDd$$

Step 3: Determine all the possible gametes each parent can produce and write the symbols for the alleles in a Punnett square.

Because there are two pairs of alleles in a double-factor cross, each gamete must contain one allele from each pair: one from the *E* pair (either *E* or *e*) and one from the *D* pair (either *D* or *d*). In this example, each parent can produce four different kinds of gametes. The four squares on the left indicate the gametes produced by the male; the four on the top indicate the gametes produced by the female. To determine the possible allele combinations in the gametes, select one allele from the "ear" pairs of alleles and match it with one allele from the "color" pair of alleles. Then match the same "ear" allele with other "color" allele. Then select the second "ear" allele and match it with each of the "color" alleles. This may be done as follows.

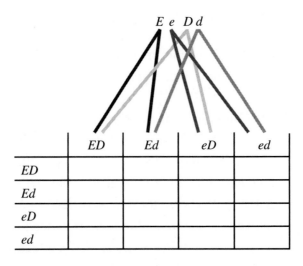

316

Step 4: Determine all the gene combinations that can result from the combining of gametes. Fill in the Punnett squares as follows:

	ED	*Ed*	*eD*	*ed*
ED	*EEDD* *	*EEDd* *	*EeDD* *	*EeDd* *
Ed	*EEDd* *	*EEdd* ^	*EeDd* *	*Eedd* ^
eD	*EeDD* *	*EeDd* *	*eeDD* "	*eeDd* "
ed	*EeDd* *	*Eedd* ^	*eeDd* "	*eedd* +

Step 5: Determine the phenotype of each possible allele combination shown in the offspring. In this double-factor problem, there are 16 possible ways in which gametes could combine to produce offspring. There are four possible phenotypes in this cross. They are represented as the following.

Genotypes	Phenotype	Symbol
EEDD, EEDd, EeDD, or *EeDe*	Free earlobes, dark hair	*
EEdd, Eedd	Free earlobes, light hair	^
eeDD, eeDd	Attached earlobes, dark hair	"
eedd	Attached earlobes, light hair	+

The probability of having a given phenotype is.

9/16 free earlobes, dark hair;
3/16 free earlobes, light hair;
3/16 attached earlobes, dark hair; and
1/16 attached earlobes, light hair.

Additional Double-Factor Inheritance Problems (Two Traits Followed from One Generation to the Next)

5. In horses, black color (B) dominates chestnut color (b). Trotting gait (T) dominates pacing gait (t). A cross is made between a horse homozygous for both black color and pacing gait and a horse homozygous for both chestnut color and trotting gait. List the probable genotype and phenotype of offspring resulting from such a cross.

Genotype _____ ; phenotype _____

6. Humans may have Rh$^+$ blood or Rh$^-$ blood. A person who is Rh$^+$ (*R*) has a certain type of protein on the red blood cell. A person who is Rh$^-$ (*r*) does not have this particular protein. In humans, Rh$^+$ dominates Rh$^-$. Normal insulin (*I*) production dominates abnormal insulin (*i*) production

(diabetes). If both parents are heterozygous for both Rh$^+$ and normal insulin production, what phenotypes would they produce in their offspring? What are the probabilities of producing each phenotype?

7. For this problem, use the information concerning the traits given in problem 2. The father is homozygous for Rh$^+$ and has diabetes. The mother is Rh$^-$ and is homozygous for normal insulin production. What phenotype would their offspring show?

8. In certain breeds of dogs two different sets of alleles determine the color pattern. Black color is dominant, and red color is recessive; solid color is dominant, and white spotting is recessive. A homozygous black-and-white spotted male is crossed with a red-and-white spotted female. What is the probability of them producing a solid black-colored puppy?

9. In humans, a type of blindness is due to a dominant allele; normal vision is the result of a recessive allele. Migraine headaches are due to a dominant allele, and normal (no headaches) is recessive. A male who is heterozygous for blindness and does not suffer from headaches marries a woman who has normal vision and does not suffer from migraines. Could they produce a child with normal vision who does not suffer from headaches? If yes, can the probability of such a child be determined?

10. This problem is a little more challenging than the previous ones because it involves lack of dominance in both characteristics that are being followed. However you use the same Punnett square method to determine the outcome of this cross as you would with any other double-factor cross.

   In the radish plant, the long and round traits exhibit lack of dominance, and the heterozygotes have an oval shape. The red and white color traits also exhibit lack of dominance, and heterozygotes have a purple color. Two oval-shaped, purple plants are crossed. What phenotypic ratio would the offspring show?

Sex-Linked Problems (Alleles Located on the X Chromosome)

   For these problems you need to remember that human males have one X chromosome and one Y chromosome; females have two X chromosomes. The Y chromosome does not carry the genes found on the X chromosome but carries genes that determine maleness.

11. In humans, the condition for normal blood clotting (*H*) dominates the condition for nonclotting (*h*) (hemophilia). Both alleles are linked to the X chromosome. A male hemophiliac marries a woman who is a carrier for this condition. (In this respect, a carrier is a woman who has an allele for normal blood clotting and an allele for hemophilia.) If they have a son, what are the chances he will be normal for blood clotting?

12. For this problem, use the information given in problem 1. A male who has normal blood clotting marries a woman who is a carrier for hemophilia. What are the chances that they will have a son who is normal for blood clotting?

13. Color blindness is a condition in which a person cannot distinguish specific colors from one another. For example they may not be able to distinguish red from green or blue from yellow. However, they are able to distinguish some colors. Because color-blind people are not blind, and they can see some colors, many people prefer to use the term color-deficient. In humans, the condition for normal vision dominates color blindness. Both alleles are linked to the X chromosome. A color-blind male marries a color-blind female. If they have a daughter, what are the chances she will have normal vision?

14. For this problem, use the information given in problem 3. A male with normal vision marries a woman who is colorblind. She gives birth to a daughter who is also colorblind. The husband claims the child is not his. The wife claims the child is his. Can you support the argument of either parent? If yes, which one? Why?

Multiple Allele Problems (Characteristics That Have More than Two Possible Forms of the Same Gene)

15. In humans, there are three alleles for blood type: A, B, and O. The allele for blood type A and the allele for blood type B show incomplete dominance. A person with both alleles has blood type AB. Both A and B dominate type O. A person with alleles for blood types A and O marries someone with alleles for blood types B and O. List the types of blood their offspring could have and the probability for each blood type in the offspring.

16. For this problem, use the information given in problem 1. A young woman with blood type O gave birth to a baby with blood type O. In a court case, she claims that a certain young man is the father of her child. The man has type A blood. Could he be the father? Can it be proven on this evidence alone that he is the father?

17. In humans, kinky hair (H^{++}), curly hair (H^{+}), wavy hair (H), and straight hair (h) are dominant, in that order. Dark hair dominates red hair. A wavy, red-haired male, whose mother has straight, dark hair, marries a dark-haired female with straight hair who has a curly, red-haired father. What type of children can they produce, and what is the probability of producing these types of offspring?

Epistasis Problems

Epistasis occurs when one set of alleles can mask the presence of a different set of alleles. All the epistasis problems in this exercise involve two different traits; therefore, they are all similar to double-factor problems.

18. Normal pigmentation dominates no pigmentation (albino). For an organism to exhibit color, it must have an allele for normal pigment production as well as alleles for a specific color. In cattle, red color dominates black color. An albino bull that has a heterozygous genotype for red is crossed with a red cow. The cow is heterozygous for normal pigment production and for red coloring. What types of offspring will they produce, and what is the probability of producing these types of offspring?

19. In humans, normal pigmentation dominates no pigmentation (albino). Black hair dominates blonde hair. An albino person will have white hair even though he or she may also have the alleles for black or blonde hair. An albino male who is homozygous for black hair marries a woman who is heterozygous for normal pigmentation and has blonde hair. What colors of hair can their children have, and what is the probability for each hair color?

Polygenic Inheritance

Polygenic inheritance occurs when two or more genes combine their effects to determine the phenotype.

20. In some types of wheat, color is caused by two sets of alleles. To produce a red color, both dominant alleles, *R* and *B*, are needed. White results from having both recessive alleles in the homozygous state, *rrbb*. Any other combination produces brown wheat grains. A strain with a genotype of *Rrbb* is crossed with a strain of wheat with a genotype of *rrBb*. What is the color of each of the parent strains?

        Rrbb color _____ ; *rrBb* color _____

What colors of wheat result from this cross, and what is the probability for each color?

Determination of Genotypes (Genetic Detective Work)

Not all heredity problems deal with determining the genotype and phenotype of the offspring. A common problem is to determine the genotype of all individuals involved when only the phenotypes are known.

In humans, free earlobes are dominant, and attached earlobes are recessive. Two free-earlobed people marry and produce one free-earlobed child and two attached-earlobed children. What are the genotypes of the parents and each of the children?

Because both parents have free earlobes, they must have at least one allele for free earlobes.

So their genotypes would be E __ (the __ means the allele could be either the dominant allele or the recessive allele). The genotype for the attached-earlobed children must be *ee*. Because each parent contributed one allele to each attached-earlobe offspring, you know that each free-earlobed parent has an allele for attached earlobes. Therefore, the __ must be *e*, and the genotype for each parent is *Ee*. All you know about the free-earlobed offspring is that the child has one allele for free earlobes. The genotype could be either *EE* or *Ee*.

In the remaining problems, try to determine the genotypes of the individuals. Take it slowly; put down an allele only when you are certain the individual has that allele. If you are not certain, show that you don't know by leaving a blank (__).

21. Normal pigmentation (*A*) dominates no pigmentation (albino = *aa*). Dark hair coloring (*D*) dominates light hair coloring (*d*). Two people with normal pigmentation produce one child with dark hair, two children with light hair, and two albino children. What are the possible genotypes for the parents?

22. A red bull, when crossed with white cows, always produces roan-colored offspring. Explain how the colors for red, white, and roan are inherited.

23. In rabbits, short hair is due to a dominant allele, *S*, and long hair to its recessive allele, *s*. Black hair is due to a dominant allele, *B*, and white hair to its recessive allele, *b*. When two rabbits are crossed, they produce 2,518 short-haired, black offspring and 817 long-haired, black offspring. What are the probable genotypes of the parents?

24. In humans, the condition for normal blood clotting dominates hemophilia. Both alleles are sex-linked to the X chromosome. Two parents produce daughters who are all carriers and sons who are all normal. What are the probable genotypes of the parents?

25. In humans, deafness is due to a homozygous condition of either or both recessive alleles *d* and *e*. Both dominant alleles *D* and *E* are needed for normal hearing. Two deaf people marry and produce offspring who all have normal hearing. What are the probable genotypes of the children and parents?

Answers

1. a. Six fingers; six fingers

   b. *Ff*; *Ff*

   c. *F* or *f*

   d. *F* or *f*

   e. 3/4

   f. 1/4

2. 1/2; 1/2

3. 1/4 red; 1/2 pink; 1/4 white

4. 1/2 pink; 1/2 red

5. *BbTt*; all black trotters

6. 9/16 Rh$^+$, normal insulin; 3/16 Rh$^+$, diabetic; 3/16 Rh$^-$, normal insulin; 1/16 Rh$^-$, diabetic

7. All Rh$^+$, normal insulin

8. None

9. Yes; 1/2 normal and no headache

10. 1/16 long, red; 2/16 long, purple; 1/16 long, white; 2/16 oval, red; 4/16 oval, purple; 2/16 oval, white; 1/16 round, red; 2/16 round, purple; 1/16 round, white

11. 1/2

12. 1/4

13. None

14. Yes. Father. A female requires two color-blind alleles to have the condition, and the father does not have any color-blind alleles.

15. 1/4 A, 1/4 B, 1/4 O, 1/4 AB

16. Yes. No. Blood typing can only disprove paternity.

17. 1/4 straight, red-haired; 1/4 wavy, red-haired; 1/4 straight, dark-haired; 1/4 wavy, dark-haired

18. 8/16 albino, 6/16 red, 2/16 black or 1/2 albino, 3/8 red, 1/8 black

19. 1/2 albino (white hair), 1/2 black hair

20. Brown; brown. 1/4 red, 1/2 brown, 1/4 white

21. *Aa_d* × *AaDd*

22. Incomplete dominance

23. *SsBB* × *Ss _ _*

24. Normal mother; hemophiliac father

25. Parents' genotype *ddEE DDee*; children's genotype *DdEd*

Experiment 38: Human Variation

Invitation to Inquiry

Historically there has been much interest in what are called racial differences. Analyze the concept of race from a purely biological point of view. Do not include differences in culture such as hair styles, speech patterns, types of family units, religious preference, or similar factors.

Does race in the human population have meaning from a purely biological point of view?

Background

Have you ever wondered why people vary so much in appearance, even when they are closely related? These differences exist because all individuals inherit unique combinations of genes from their parents. This variety is possible because there are often many different **alleles** (alternative forms of a gene) for a specific characteristic within a population, and they are mixed in new combinations during sexual reproduction. Each child receives one half of its genetic information from each parent. More accurately, each parent contributes one allele for each genetic **locus** (the specific location of a gene on a chromosome). However, because of meiosis even children of the same parents will receive a different set of genes from each parent. (Identical twins are an exception because they are the result of a single fertilization event.)

Several different patterns of inheritance have been documented. Because each individual gets one allele for each characteristic from each parent, the two alleles must either be the same or different. When both alleles for a characteristic are identical, the individual is said to be **homozygous**. If the two alleles for a characteristic are different, the individual is said to be **heterozygous**. Many alleles are dominant, which means they mask or hide the expression of recessive alleles. Put another way, **dominant** alleles are expressed, even when present in just a single dose. **Recessive** alleles can be expressed only when dominant alleles are absent. In some cases, two alleles may not show dominance or recessiveness, and both express themselves, a condition known as **incomplete dominance** (lack of dominance). Often such heterozygous cases result in phenotypic characteristics that are intermediate between the two homozygous genotypes. Many characteristics have only two alleles available in the population. However, many cases of **multiple alleles** exist in which there are many alleles for a characteristic within a population although an individual can only have a maximum of two of the alleles, one having been inherited from each parent. Other traits, such as beard growth, are **sex-limited**, which means their expression is limited to one sex or the other.

Other human traits are not determined by a single set of alleles and show a wide variety of phenotypic expressions among the individuals in a population. These traits are thought to be controlled by many sets of alleles (genes) that are located at different loci, and these traits are called **polygenic characteristics**. Some traits are influenced by the action of genes at two or more loci. For example, the alleles that determine hair color or skin color are affected by another set of alleles that determines whether a person can produce pigment or is an albino. This phenomenon, genes at one locus influencing how the genes at another locus are expressed, is what geneticists call **epistasis**. This laboratory activity provides an opportunity to observe how alleles are passed from one generation to the next and how combinations of alleles determine the phenotype of offspring. This exercise will also illustrate how each of the topics discussed earlier (dominant-recessive, incomplete dominance, multiple alleles, sex-limited, polygenic, and epistasis) influences the pattern of inheritance.

In this exercise you will choose a partner of the opposite sex. The two of you will be the parents of an offspring. (If there are unequal numbers of males and females in the class, your instructor may designate you to be a particular sex for the purposes of this laboratory.) "Parents" will be assumed to be heterozygous for characteristics and will flip coins to simulate segregation of alleles during the formation of gametes. Alleles that are on different chromosomes are segregated independently of one another. This **independent assortment** of chromosomes during the formation of gametes will be simulated by additional flips of coins. We will use facial characteristics in this exercise, and you will draw a picture of what the child would look like in his or her teens.

The exact manner in which human facial characteristics are determined is often difficult to determine. Some of the characteristics are inherited in exactly the manner suggested and are indicated with a footnote. Others are assumed to be inherited in the manner suggested and are indicated with an asterisk (*). Others are useful for the laboratory exercise and are assumed to be inherited, but the manner of inheritance is not understood; these characteristics will not have a footnote or asterisk.

In this lab exercise you will
1. flip coins to determine which alleles are passed to your offspring and determine the genotype of your offspring.
2. record your data on the data sheet.
3. make a drawing of your offspring's phenotype based on the genotype obtained.
4. share your data with the class and compute the genotypic and phenotypic ratios for certain characteristics using the data from the entire class.
5. examine different modes of inheritance including simple dominance and recessiveness; incomplete dominance; multiple alleles and sex-limited, polygenic, and epistatic traits.
6. compute a gene frequency from a set of data.
7. use chi-square analysis to determine if deviation from the expected Mendelian ratios could be due to chance alone.

Procedure

Pair up with a classmate who will play the role of your spouse. Begin the simulation with the assumption that each of you has one dominant and one recessive allele for each of the facial features illustrated on the following pages. In other words, each of you is *heterozygous* for each trait. To determine which allele you will pass on to your child, both you and your spouse flip a coin. "Heads" determines that a dominant allele is present in the gamete and is passed on to the offspring. "Tails" determines that a recessive allele is present in the gamete and is passed to the offspring. Thus, if your partner flips heads and you flip tails, the child's genotype for that trait would be heterozygous (*Aa*). If you both flip heads, the child's genotype would be *AA*, and if you both flip tails, the child's genotype would be *aa*.

1. First we should determine the sex of the child. Females have two X chromosomes and males have an X and a Y chromosome. Which parent determines the sex of the child? All egg cells have X chromosomes, but half of the sperm have Y chromosomes and half have X chromosomes. If a sperm cell bearing a Y chromosome fertilizes an egg, a male (XY child) will result; if a sperm bearing an X chromosome fertilizes an egg, a female (XX child) will result. Therefore, in this simulation, only the father needs to flip the coin to determine the sex of the child. If heads is flipped, your child is a boy (Y-bearing sperm), and if tails is flipped, your child is a girl (X-bearing sperm).

2. Give your child a name and record this information on the data sheet on page 339.

3. Both you and your partner should flip a coin for each facial feature. Record (a) the alleles contributed by each parent, (b) the genotype of the offspring, and (c) the phenotype of the offspring on the data sheet on page 337.

Facial Traits

1. Face shape: *RR* or *Rr* = round face; *rr* = square face.

Round face (*RR*, *Rr*)

Square face (*rr*)

2. Chin size: *PP* or *Pp* = very prominent chin; *pp* = less prominent chin.

Very prominent chin (*PP*, *Pp*)

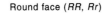

Less prominent chin (*pp*)

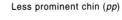

3. Chin shape: Only flip coins for this trait if chin size genotype is *PP Pp*. The genotype *pp* prevents the expression of chin shape. *CC* or *Cc* = round chin; *cc* = square chin.

Round chin (*CC, Cc*)   Square chin (*cc*)

4. Cleft chin: 11900[1] *AA*, or *Aa* = cleft chin; *aa* = no cleft chin.

Present (*AA, Aa*)   Absent (*aa*)

Skin Color

5. Skin color*: To determine skin color, assume there are three sets of alleles, located at three different loci, that contribute to the amount of pigment produced. Both you and your partner should flip coins to determine the genotype of the first pair of alleles (*AA, Aa, aa*). Then flip again to determine the genotype of the second pair of alleles (*BB, Bb, bb*). Flip for the last time to determine the third pair of alleles (*CC, Cc, cc*). Determine the phenotype of your offspring based on the following polygenic model.

> Six dominant alleles—very dark black
> Five dominant alleles—very dark brown
> Four dominant alleles—dark brown
> Three dominant alleles—medium brown
> Two dominant alleles—light brown
> One dominant allele—light tan
> Zero dominant alleles—fair skin

Example: If you flipped heads for the first two sets of alleles and tails for the third set of alleles, and your partner flipped tails for the first set of alleles and heads for the second and third sets of alleles, your offspring's genotype would be *AaBBCc,* and your offspring's phenotype would be dark brown.

[1] Reference number to Victor A. McKusick, M.D., *Mendelian Inheritance in Man*, 8th ed. Baltimore and London: Johns Hopkins University, 1988.

Hair Traits—Next Four Flips

6. White forelock*: A tuft of hair over the forehead is white.

FF, or *Ff* = white forelock; *ff* = no white forelock

7. Widow's peak*: The hairline comes to a point in the center of the forehead.

WW, or *Ww* = widow's peak present; *ww* = widow's peak absent

Present (*WW, Ww*)   Absent (*ww*)

8. Hair type and hair color*: Hair type and hair color are determined by multiple alleles. To determine which alleles you have and are able to contribute to your offspring your instructor will have designated each individual in class as belonging to one of five groups. From the "Special Characteristics Chart" determine your genotype and which alleles you may pass to your offspring. The alleles for hair shape are S^K=kinky, S^C = curly, S^W = wavy, and S^{ST} = straight. They exhibit dominance and recessiveness in the order given. Kinky is dominant to all other alleles, straight is recessive to all other alleles. Curly is recessive to kinky but dominant to wavy. Wavy is recessive to kinky and curly and dominant to straight.

The alleles for hair color are C^{BK} = black, C^{BR} = brown, C^{BD} = blond, and C^R = red. They exhibit dominance and recessiveness in the order given. Determine the hair shape and hair color of your offspring by flipping coins.

Special Characteristics Chart			
Group Assigned by Instructor	Hair Type Genotype	Hair Color Genotype	Albinism Genotype
Group 1	$S^K S^C$	$C^{BK} C^{BR}$	*aa*
Group 2	$S^K S^{ST}$	$C^{BK} C^R$	*AA*
Group 3	$S^C S^W$	$C^{BR} C^{BD}$	*AA*
Group 4	$S^C S^{ST}$	$C^{BR} C^R$	*Aa*
Group 5	$S^W S^{ST}$	$C^{BD} C^R$	*Aa*

331

Once you have determined your genotype for hair type and color, complete the following chart.

Trait	My Genetic Information		My Partner's Genetic Information	
	Genotype	Phenotype	Genotype	Phenotype
8a. Hair Type				
8b. Hair Color				

Now flip your coins to determine which hair type and hair color alleles each of you will pass on to your offspring.

9. Albinism: Albinos are not able to produce pigment. Therefore none of the skin color, hair color, or eye color genes can express themselves if an individual is an albino. The recessive gene for albinism is rare in the population. To determine your genotype for the ability to produce pigment, refer to the Special Characteristics Chart.

Once you have determined your genotype and phenotype, complete the following chart.

Trait	My Genetic Information		My Partner's Genetic Information	
	Genotype	Phenotype	Genotype	Phenotype
Pigment production				

Determine if your child is an albino by flipping coins based on the genotype you were assigned. If you are homozygous, you do not need to flip a coin because you can only pass on one type of allele. If your child is an albino, all color characteristics will be altered: eyes will be pink because there is no pigment in the iris, the skin will be very pale white/pink because there is no pigment, and all hair (on head, eyebrows, eyelashes, etc.) will be white regardless of the other color alleles he/she may have inherited. Also white forelock will not show because all hair will be white, and freckles will not show because no pigment is produced in the skin. Albinism is inherited in the following manner:

AA or Aa = normal pigment; aa = no pigment (albino).

Eyebrow Traits—Next Two Flips

10. Eyebrow thickness: BB = very bushy; Bb = intermediate; bb = very thin.

Bushy (*BB*)   Intermediate (*Bb*)   Very thin (*bb*)

11. Eyebrow placement: *NN* or *Nn* = not connected; *nn* = connected

Not connected (*NN, Nn*)   Connected (*nn*)

Eye Traits

12. Eye color: Dark brown eyes are produced by the presence of a layer of brown pigment that covers the entire surface of the iris. Lighter eyes (hazel) are caused by patches of brown pigment on the iris. Blue eyes are caused by the absence of pigment on the iris. In this situation, the dominant allele *S* is for solid pigment and the recessive allele *s* is for patchy pigment. The dominant allele *B* is for the presence of brown pigment and the recessive allele *b* is for the absence of brown pigment.

To determine eye color, both you and your partner will need to flip twice. Determine the genotype of the first pair of alleles (*SS, Ss, ss*) and then the second pair of alleles (*BB, Bb, bb*). Determine the phenotype of your offspring according to the following guidelines.

SSBB, SSBb, SsBB, or *SsBb*	brown
ssBb or *ssBB*	hazel
SSbb, Ssbb or *ssbb*	blue

13. Eye distance: *EE* = close together; *Ee* = intermediate; *ee* = far apart.

Close together (*EE*)   Average distance (*Ee*)   Far apart (*ee*)

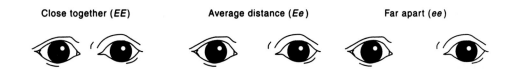

14. Eye shape: *AA* or *Aa* = almond shaped; *aa* = round.

Almond (*AA, Aa*)   Round (*aa*)

15. Eyelashes: *LL* or *Ll* = long; *ll* = short.

Long (*LL, Ll*)   Short (*ll*)

Mouth and Lip Traits

Determine the phenotype with respect to all three characteristics before drawing the mouth.

16. Mouth size: *MM* = wide; *Mm* = intermediate; *mm* = narrow.

Long (*MM*) Average (*Mm*) Short (*mm*)

17. Lip thickness: *TT* or *Tt* = Thick; *tt* = thin.

Thick (*TT*, *Tt*) Thin (*tt*)

18. Dimples on side of mouth: *DD* or *Dd* = dimples; *dd* = no dimples.

Present (*DD*, *Dd*) Absent (*dd*)

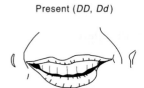
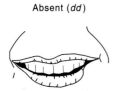

Nose Traits

19. Nose thickness: *BB* or *Bb* = broad nose; *bb* = narrow nose.

Broad (*BB*, *Bb*)

Narrow (*bb*)

20. Nose shape: *RR* or *Rr* = tip rounded; *rr* = tip pointed.

Pointed (*rr*)

Rounded (*RR*,*Rr*)

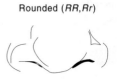
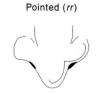

Ear Traits

21. Earlobe attachment: *FF* or *Ff* = free earlobes; *ff* = attached earlobes.

Free (*FF*, *Ff*)

Attached (*ff*)

22. Darwin's earpoint: *DD* or *Dd* = point present; *dd* = point absent.

Present (*DD*, *Dd*)

Absent (*dd*)

23. Ear pits: 12870[2] *PP* or *Pp* = pits present; *pp* = pits absent.

Present (*PP*, *Pp*)

Absent (*pp*)

24. Hairy ears: Hairy ears is a recessive allele located on the Y chromosome, so it is only contributed by the male and only males can display the characteristic. If your child is female it cannot show the characteristic, so you do not need to do anything. If your child is a male the father must flip a coin to decide if the child will have hairy ears. Tails denotes hairy ears.

Absent

Present

---

[2] Reference number to Victor A. McKusick, M.D., *Mendelian Inheritance in Man*, 8th ed. Baltimore and London: Johns Hopkins University, 1988.

Freckles

25. Freckles on cheeks: *CC* or *Cc* = freckles present; *cc* = freckles absent.

Present (*CC, Cc*)

Absent (*cc*)

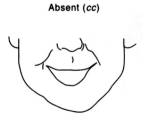

26. Freckles on forehead: *FF* or *Ff* = freckles present; *ff* = freckles absent.

Present (*FF, Ff*)

Absent (*ff*)

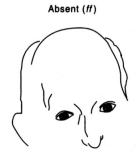

Data Sheet

Parents' names _____ and _____

Child's Name _____

Child's Sex _____

Trait no. Trait	Gene from mother	Gene from father	Genotype	Phenotype
1 Face shape				
2 Chin size				
3 Chin shape				
4 Cleft chin				
5 Skin color				
6 White forelock				
7 Widow's peak				
8a Hair type				
8b Hair color				
9 Albinism				
10 Eyebrow thickness				
11 Eyebrow placement				
12 Eye color				
13 Eye distance				
14 Eye shape				
15 Eyelashes				
16 Mouth size				
17 Lip thickness				
18 Dimples				
19 Nose thickness				
20 Nose shape				
21 Earlobe attachment				
22 Darwin's earpoint				
23 Ear pits				
24 Hairy ears				
25 Freckles on cheeks				
26 Freckles on forehead				

Name _____ Lab section_____

Your instructor may collect these end-of-exercise questions. If so, please fill in your name and lab section.

End-of-Exercise Questions

1. Compare your child with the child of another couple. List five traits that both of the children have. The children are concordant for these traits.

2. Using the same children for comparison, list five traits that are different. The children are disconcordant for these traits.

3. For most of the characteristics in this exercise both parents are heterozygous. What is the probability that both parents will contribute a recessive allele for any given trait?

4. Refer to your child's data sheet and complete table 38.1. Place an X in the dominant column in table 38.1 if your child received at least one copy of the dominant allele. Place an X in the recessive column if your child received two copies of the recessive allele. Total the number of dominant and recessive phenotypes for these traits. What is the ratio of dominant-to-recessive traits? What ratio would you expect? Why?

Table 38.1	Ratio of Dominant to Recessive Phenotypes	
Trait	Dominant	Recessive
1 Face shape		
2 Chin size		
4 Cleft chin		
6 White forelock		
7 Widow's peak		
11 Eyebrow placement		
14 Eye shape		
15 Eyelashes		
17 Lip thickness		
18 Dimples		
19 Nose thickness		
20 Nose shape		
21 Earlobe attachment		
22 Darwin's earpoint		
23 Ear pits		
25 Freckles on cheeks		
26 Freckles on forehead		
Total		

5. There are three characteristics that demonstrate incomplete dominance in this exercise. What are they?

6. If the genes for earlobe shape and dimples were located close to one another on the same chromosome, how would their location influence how these two genes are passed to the next generation?

7. List two examples of epistasis described in this exercise.

8. What type of inheritance pattern best describes how skin color is determined? Ignore the cases of albinism.

9. Analyze the number of dominant versus recessive phenotypes recorded in table 38.1. Do your results agree with a Mendelian mode of inheritance? Use a chi-square test to determine if the deviation from the expected results could be attributed to chance. Make up your own chi-square table.

10. What impact do cases of multiple alleles have on the number of kinds of phenotypes displayed in the population?

Experiment 39: Sensory Abilities

Invitation to Inquiry

There are many kinds of eyeglasses used for special purposes. People who fish like to wear polarizing sunglasses. People who shoot guns competitively typically wear amber colored glasses. Conduct some research to determine why each prefers a particular kind of eyewear.

Background

This laboratory exercise gives you an opportunity to study how we sense changes in our surroundings. Your ability to sense changes in your surroundings involves (1) the specific ability of sense organs to respond to stimuli (detection), (2) the transportation of information from the sense organ to the brain by way of the nervous system (transmission), and (3) the decoding and interpretation of the information by the brain (perception). In order for us to sense something, all three of these links must be functioning properly. For example, a deaf person might be unable to detect sound because (1) there is something wrong with the ear itself, (2) the nerves that carry information from the ear to the brain are damaged, or (3) the portion of the brain that interprets information about sound is not functioning properly. While this laboratory activity focuses on the function of sense organs, it is important to keep in mind that the peripheral and central nervous systems are also important in determining your sensory ability. All sense organs contains specialized cells that are altered in some way by changes in their environment (stimuli). The sensory cells depolarize, and since they are connected to nerve cells, they cause the nerve cells to which they are attached to depolarize as well, and information is sent to the brain for interpretation by way of nerve pathways.

In this lab exercise you will
1. make a map of the location of different kinds of taste buds on your tongue.
2. determine several characteristics of the sense of "touch."
3. locate different kinds of temperature sensors in the skin.
4. study several aspects of visual acuity.
5. study several aspects of the sense of hearing.

Procedure

Taste

Taste involves several different kinds of sensory cells located on the tongue and pharynx. Each kind of sensory cell responds to specific kinds of chemicals. So there is not just one sense of taste; there are several. We recognize at least five different kinds of taste senses: sweet, sour, salty, bitter, and umami (meaty).

Mapping the Sense of Taste on the Tongue

1. Work with a lab partner.
2. Obtain a cotton swab and dip it into one of the solutions. The solutions are labeled sweet, sour, salt, bitter, and umami (meaty).
3. Have your lab partner touch the swab to the tongue at the following five locations: *a*. the tip, *b*. right side, *c*. left side, *d*. center, and *e*. back.

Place an X on the following drawings of the tongue to indicate where you detected each chemical.

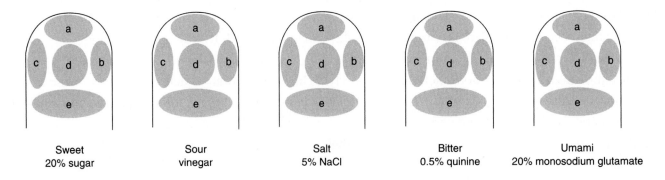

Sweet	Sour	Salt	Bitter	Umami
20% sugar	vinegar	5% NaCl	0.5% quinine	20% monosodium glutamate

4. Test the other four solutions in the same manner, but be sure to rinse your mouth with water after each solution.
5. When you have tested each of five chemicals, switch positions with your partner.

Results

1. Can you detect each chemical at all places on the tongue?

2. Compare your results to your partner and other people in class. Do they detect the same chemicals in the same place?

3. What does this tell you about the sense of taste?

The Role of Solubility in Detecting Taste

1. Dry off the tip of your tongue with a clean paper towel. Place a few grains of table salt (NaCl) on the tip of your tongue. Record the time interval from the time you place salt on the tip of your tongue until you first taste the salt. _____
2. Dissolve a few grains of salt in a small amount of water. Place this on the tip of your tongue. Record the time interval from the time you place the salt solution on the tip of your tongue until you first taste the salt. _____
Were the two time intervals different? What does this tell you about the ability to taste salty materials?

Touch

The sense of touch is made up of a number of different types of receptor organs. Pressure, pain, heat, and cold are all aspects of the sense of touch. We will experiment with some of them here.

Localization of Touch

You need a partner for this exercise.

1. The subject should keep his or her eyes closed throughout the exercise.
2. Touch the skin of the back of the hand of the subject lightly with the pointed end of a soft lead pencil. Be sure to leave a mark.
3. Then ask the subject (with eyes still closed) to use a blunt probe to locate the place on the skin

where the stimulus was received.

4. Use a ruler to measure as closely as possible the error in locating where the stimulus was applied. Measure the error in millimeters. Repeat five times at different locations on the back of the hand.

5. Change roles with your partner and repeat the experiment.

Results and Conclusions

In the space provided, write a short paragraph that states your findings and conclusions.

Density of Sense Organs

You need to work in pairs.

1. Have the subject keep his or her eyes closed.

2. Use a pair of forceps or calipers to gently touch the subject's skin so that the two points of the instrument touch with the same light pressure and at the same time. Test the palm of the hand and two other regions of the body. Other regions that may be tested are the back of the hand, the tip of the index finger, the forearm, the tip of the nose, the forehead, and the back of the neck. Not all of these need to be tried, but a selection should be made.

3. Ask the subject to state whether one or two points of the instrument are felt. Repeat this procedure five times for each area of the body chosen. (To keep the subject from guessing, the experimenter should occasionally touch the skin with only one point. However, do not record the result of the response in your data).

4. Record your data in the following manner: Record a minus sign (–) whenever two points were felt as one and a plus sign (+) whenever the two points were actually felt as two.

5. Begin with the points 20 millimeters apart and systematically decrease the distance between the points from 20 mm to 15 mm to 10 mm to 5 mm. Find the smallest distance at which the subject can still distinguish two points for each portion of the body tested.

6. Change roles. Record the data made on yourself as the subject.

7. From the data, estimate the comparative densities of touch receptors of the different parts of the body.

Area I: _____

distance between points of forceps in mm

	Trial Number				
	1	2	3	4	5
20 mm					
15 mm					
10 mm					
5 mm					

Area II: _____

distance between points of forceps in mm

	Trial Number				
	1	2	3	4	5
20 mm					
15 mm					
10 mm					
5 mm					

Area III: _____

distance between points of forceps in mm

	Trial Number				
	1	2	3	4	5
20 mm					
15 mm					
10 mm					
5 mm					

Results and Conclusions

1. What is the smallest distance the subject can still recognize two points for each of the three areas tested? Are they the same? Explain.

2. Place a sketch of your "two-point device" on the drawing to indicate why two points are sometimes felt as one.

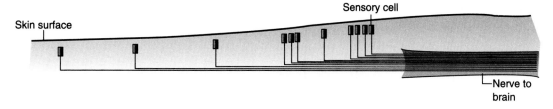

3. Which of the regions of the skin that you tested is represented by the left side of the drawing, and which is represented by the right side of the drawing? Explain your answer.

Temperature Sense–Detecting Hot and Cold

Work with a partner.

1. With a pen, draw a square with 20 mm sides on the back of the subject's hand, then subdivide this square into 16 smaller squares by dividing each of the sides into 5 mm segments.
2. Have the subject keep eyes closed and place his or her hand flat on the table.
3. Obtain a nail that has been in ice-cold water.
   Dry it off with a paper towel.
   Lightly touch each of the squares of the grid on the hand at random.
   The subject should respond by saying "cold" if such a sensation is actually felt; otherwise the subject remains silent. *It is important for the subject to ignore the sense of touch and concentrate on the sensation of cold.*
4. For every positive response, the experimenter marks a plus sign (+) on the following grid at a point corresponding to the point tested on the skin.
5. Be sure that the nail is really cold when you make each test.
6. Repeat this exercise with a very warm nail and record your results on the second grid.

Cold					Warm			

7. Switch roles with your partner and repeat the exercise.
8. Answer the following questions.

   Do you detect hot in every square?_____

   Do you detect cold in every square?_____

   Are hot and cold receptors always located in the same squares?_____

Do the same receptors respond to hot and cold? Explain how you know.

Temperature Sense—Detecting Changes in Temperature

1. Dip one finger into a beaker of hot water and at the same time put a finger from the other hand into cold water.
2. After 30 seconds, transfer both fingers into a third beaker of warm water.

Results and Conclusions

Describe the sensations of both fingers in the beaker of warm water and explain why there is a difference in sensation.

Vision

The eye is a complex structure that focuses light on cells of the retina that respond to changes in light. There are two kinds of light receptors, rods and cones. Rods are very sensitive to light and only respond to differences in light intensity. The cones are less sensitive to light. There are at least three kinds of cones, each of which responds to specific colors of light. The rods and cones are located in different places in the retina of the eye. In this part of the lab activity you will make a number of observations about the eyes and their response to various stimuli.

Light Intensity and Color Vision

Work with a partner.

1a. Take three pieces of different but similarly colored paper that are about 100×100 cm into a nearly dark room. Show only one square at a time and ask your partner to identify the color of the paper. Determine the distance at which your partner can tell the color of the squares of paper. It is not necessary to measure the distance exactly. Simply count the number of paces between you and your partner.
1b. Change roles and have your partner show you the papers.
2. Return to a well-lighted area and determine the distance at which your partner can still identify the colors.

Explain your results by discussing the function of the rods and cones in the retina.

Determining the Location of Rods and Cones

Rods and cones are not located in the same place on the retina of the eye. When you look at things from directly in front of the eye, the cornea and lens of the eye focus the light on a region known as the fovea centralis. When you look at things with your peripheral vision, the light is focused on regions of the eye other than the fovea centralis.

Work with a partner.
1. Choose three similarly colored squares of paper about 100×100 cm.
2. Have your partner stare at a distant object directly in front of him or her.
3. Start behind your partner (out of the field of vision) and slowly move the piece of paper forward at eye level about 30 cm to the side of the head.
4. Ask your partner to tell you when the piece of paper is first seen and when the color of the paper can be detected.
5. Use the information about the location of rods and cones and the results you just obtained to answer the following questions.
   Which sense organs (rods or cones) are most common in regions outside the fovea centralis?

   _____

   Which sense organs (rods or cones) are most common within the region of the fovea centralis?_____
   Explain how this experiment allows you to answer these questions.

Detecting the Blind Spot

Use the + and dot below in the following manner. Close your left eye. Place the page close to your face. Stare at the + with your right eye. Slowly move the page away from you. What happens to the dot?

+   ●

In order to detect the presence of an object, light must fall on the retina of the eye and stimulate either rods or cones. There are no rods or cones at the point where the optic nerve goes out of the back of the eye. Use this information to explain what you observed when looking at the + and •
above.

350

Hearing

The sense of hearing involves the detection of sound vibrations. Airborne sounds cause the eardrum to vibrate. The eardrum is attached to a series of three small bones: the malleus, incus, and stapes. The stapes is attached to a membrane over a small opening in the cochlea. The cochlea is fluid filled. Thus the vibrations of the air are transferred to the fluid of the cochlea. When the fluid in the cochlea vibrates, cells in the cochlea are stimulated. When these cells depolarize, they send a signal by way of the auditory nerve to the brain. In this part of the lab activity we will explore some aspects of hearing.

Work with a partner.
1. Strike a low frequency tuning fork (100 cps) and hold it near one ear. Determine how far from the ear the subject can hear the tuning fork. Repeat with the other ear. Are both ears the same?

2. Strike the tuning fork and touch the *base* of the vibrating tuning fork to the skull just in front of the ear. Does the volume change?

How is this sensation of hearing different from when the base of the turning fork touches the skull near the ear?

3a. Have the subject sit with closed eyes.
   Strike the tuning fork.
   Have the subject point to the position of the tuning fork.
   Repeat three times from different positions.
   Can the subject correctly identify the position of the tuning fork?
3b. Now have the subject keep eyes closed and plug one ear with a finger.
   Have the subject point to the tuning fork as it is struck at different positions.
   Was the subject able to locate the position of the tuning fork accurately?
   Why was there a difference between the two different trials?

Sensory Abilities

Name _____ Lab section_____

Your instructor may collect these end-of-exercise questions. If so, please fill in your name and lab section.

End-of-Exercise Questions

1. Describe the regions of your tongue that are most sensitive to sweet, sour, salt, bitter, and umami.

2. How is solubility important to the sense of taste?

3. Determine the average distance between points on the palm of the hand at which persons in the class correctly identify that they were being touched by two points. On the average, individuals could discriminate between two points that were _____ mm apart.

4. Using the data you collected for different parts of the skin, rank them according to which had the greatest density of touch receptors and which had the lowest density.

5. Write a paragraph describing what you learned about the receptors that respond to temperature. How many kinds of receptors are there? Explain how you know there are different kinds of receptors.

6. There are some kinds of people who can see well in bright light but are not able to see in dim light. This condition is called "night blindness." What kinds of sensory cells do not function to capacity in individuals who have night blindness?

Experiment 40: Daily Energy Balance

Invitation to Inquiry

Many kinds of foods are marketed to those who participate in various kinds of sports. The implication is that these items have additional nutrients or higher quantities of nutrients needed by the athlete. Go to a store and read the ingredient label on one of these products. Compare it to an equivalent product that is not marketed in such a manner. For example, you could compare a sports drink to a soft drink or orange juice. You could compare a "nutrient bar" to an equivalent candy bar or snack food. Look specifically at quantities of calories, fats, proteins, sodium, and potassium. How different are they? What other foods could you eat that would provide the same calories and nutrients.

Background

The theoretical biological sciences of biochemistry, anatomy, cell biology, and physiology are brought together in the practical biological field of nutrition. The science of **nutrition** is the study of the processes involved in taking in, assimilating, and utilizing nutrients. The amount of food and drink consumed by a person from day to day is a person's **diet**. There has been an increased interest in diet and personal nutrition as more information concerning these subjects becomes available through the popular press, scientific publications, health clubs, and schools. Not only are people "counting calories" and concerned with the grams of fat they consume, but they are also becoming scientifically literate enough to ask significant questions of their physicians, teachers, food manufacturers, and government officials. With a minimal amount of nutritional information, it is possible to get a better handle on your own nutritional status.

In this exercise, you determine your daily basal metabolic rate, voluntary muscular activity, and specific dynamic action per day. These are used to estimate your total energy requirements per day in kilocalories (kcal). You then calculate your total daily kilocalorie intake. By comparing these two figures, you can determine whether or not your present diet should result in your maintaining, losing, or gaining weight.

You will determine your

        1. basal metabolic rate;

        2. voluntary muscular activity level;

        3. specific dynamic action;

        4. kilocalorie intake per day by adding your BMR, activity level, and SDA;

        5. total energy requirements compared to your energy kcal intake per day; and

        6. energy balance.

Procedure

Determining Your Basal Metabolic Rate

Your **BMR (basal metabolic rate)** is the rate at which kilocalories are used for maintenance activities and can be measured on a daily basis. This is also the total amount of energy per kilogram per hour expended after a 12 hour fast. Energy is measured in **kilocalories**, the amount of energy needed to raise the temperature of 1 kg of water 1°C. BMR can be estimated by using a short formula that is based on 1.0 kcal per kilogram of body weight per hour for men, or 0.9 kcal per kilogram of body weight for women. Even though this is a crude method, it does give some idea of the BMR.

Body weight × BMR factor = Estimated BMR (kcal/kg/hour)

For example, if a male weighs 150 lbs, his mass in kilograms will be 68 kg. Therefore, the estimated BMR is

68 kg × 1.0 kcal/kg/hr = 68 kcal/kg/hr,

24 hours/day × Estimated BMR/hour = Estimated energy expenditure/day,

68 kcal/kg/hr × 24 hours = 1632 kcal/kg/day.

If a female weighs 120 lbs, her mass in kilograms will be 55 kg. Therefore, the estimated BMR is

55 kg × 0.9 kcal/kg/hr = 49 kcal/kg/hr,

24 hours/day × Estimated BMR/hour = Estimated energy expenditure/day,

49 kcal/kg/hr × 24 hours = 1176 kcal/kg/day.

These are estimated basal metabolic rates for these two people. Using this method, calculate your own BMR.

Body weight in kg × BMR factor in kcal/kg/hr = Estimated energy expenditure in kcal/kg/hr

24 hours/day × Your estimated energy expenditure/hour =

	kcal/day

(your estimated energy
expenditure/day
or kcal/day used while at rest)

For a more accurate determination of your BMR, use standard tables from your text and calculate your skin surface area from your height and weight. A table of kilocalories per day per square meter of skin lists the kilocalories expended by a female or male by age group. This kilocalorie figure should be multiplied by your skin surface area to determine your BMR more accurately.

Skin surface area $\times$ Kilocalories per day per square meter of skin = BMR

Your skin surface area $\times$ Kilocalories per day per square meter of your skin =

	kcal/day

Estimating Your Energy Output per Day

Energy output per day is an estimate of your voluntary muscular activity per day. For a person who engages in only sedentary activities such as desk work, the estimated energy output is approximately 50% of his or her already determined BMR. For example, if the male in the previous example were a typist, his voluntary muscular activity level for the day would be

$$0.50 \times 1632 \text{ kcal/day} = 816 \text{ kcal/day}.$$

For a person who engages in light activities such as standing, talking, and minor amounts of walking, the estimated energy output is approximately 60% of his or her already determined BMR. For example, if the female in the previous example were a teacher, her voluntary muscular activity level for the day would be

$$0.60 \times 1176 \text{ kcal/day} = 706 \text{ kcal/day}.$$

For a person who engages in moderate activities that exceed those described as light, the estimated energy output is approximately 70% of his or her already determined BMR. For example, if the male described were a nurse, his voluntary muscular activity level for the day would be

$$0.70 \times 1632 \text{ kcal/day} = 800 \text{ kcal/day}.$$

Those participating in heavy activities are estimated to use an equivalent of their BMR per day. For a person engaged in heavy lifting and moving or a daily workout of an hour or more, his or her voluntary muscular activity level for the day would be

$$1.00 \times 1632 \text{ kcal/day} = 1,632 \text{ kcal/day}.$$

Estimate your voluntary muscular energy expenditure per day.

Percent of BMR based on activity level $\times$ BMR =

	kcal/day

(your voluntary muscular energy expenditure)

Estimating Your Specific Dynamic Action (SDA)

The specific dynamic action (SDA) is the amount of energy needed to metabolize food for the day. This is approximately 10% of a person's total daily basal expenditure and total daily voluntary muscular activity. For example, the male nurse had a voluntary muscular activity level for the day of 800 kcal, and his daily expenditure was 1,632 kcal. Therefore, his SDA for the day would be an estimated 243 kcal.

(Your basal energy expenditure + your voluntary daily physical activity) $\times$ 0.10 =

kcal/day

(your SDA)

Add together your BMR and your voluntary muscular energy expenditure.

BMR + Voluntary muscular activity =

kcal/day

Total Daily Energy Requirements

Your total daily energy requirements are the sum
of your BMR + Voluntary muscular activity + SDA =

kcal/day

Determining Your Daily Caloric Intake

Before you can draw any conclusions about your daily energy balance, you must determine your total energy intake. This can be estimated by recording your total consumption of nutrients and determining their kilocalorie values. Fill in Table 40.1 beginning with the first meal of your day and ending with the last snack you consume. The estimate of your kilocalorie intake can be determined by using the tables found in many supermarkets, bookstores, and libraries. Such tables are usually referred to as pocket calorie counters.

Compare your total daily energy requirement with your actual kilocalorie intake per day.

Total daily energy requirement – Actual kilocalorie intake = Gain/maintenance/loss

If these two figures are the same, you are meeting your energy requirements and maintaining your weight. If your total daily caloric intake is greater than your caloric requirement, you are exceeding your energy requirements and, therefore, gaining weight. The opposite is true if you are not meeting your caloric requirement. Therefore, you are losing weight.

Table 40.1	Total Energy Intake	
Food	Food serving size	Energy kcal
Breakfast		
Lunch		
Dinner		
Other		

Total = _____

Determining Your Daily Energy Balance

Name _____ Lab section_____

Your instructor may collect these end-of-exercise questions. If so, please fill in your name and lab section.

End-of-Exercise Questions

1. What is basal metabolic rate?

2. Do basal metabolic rates differ between males and females? On what evidence can you base your answer?

3. What factors are involved in accurately determining your BMR?

4. What is specific dynamic action?

5. If your total kilocalorie intake per day is higher than your total kilocalorie requirements, what happens to your weight?

6. What can you do to bring about an energy balance?

7. Other than kilocalories, what information is important in determining whether or not you are consuming a healthy diet?

8. What resources are available to help you develop a balanced diet?

Experiment 41: The Effect of Abiotic Factors on Habitat Preference

Invitation to Inquiry

In the workplace there are many kinds of abiotic factors known to affect the productivity of employees. Assume you are a manager of a business. Identify five important abiotic factors that would affect your employees' performance on the job. If you only had enough money to modify one of these factors, which one would you choose and why would you choose it?

Background

The individuals within a population are able to detect and respond to certain features of their environment. Many characteristics of a **habitat** (the space an organism inhabits) are **variable** from time to time or at different locations within the habitat. **Abiotic factors** are physical factors such as *temperature*, quantity of *light*, *gravity*, and *pH*. These often vary in aquatic habitats. When a specific environmental factor varies continuously over a distance, a **gradient** exists. Light intensities can range from absolute darkness to extreme brightness. A shady spot may be a few degrees cooler than a position in direct sunlight only a few meters away. The pH of a lake or stream may also vary from place to place. When a gradient exists it is possible for an animal to detect when the stimulus is getting stronger and either move toward or away from the stimulus. For example, if you hear a sound and want to go toward the sound, you can walk in a particular direction. If the sound is getting louder, you continue walking in that direction. However, if the sound is getting fainter, you would change your direction until it did get louder. In this manner you could follow a sound gradient to its source. It seems logical to expect that certain abiotic conditions would be more suitable for an organism to thrive and that organisms would migrate to places where the abiotic conditions are most favorable.

If we want to determine the significance of a specific variable, we need to isolate it from other variables. Then we can present a population of organisms with a gradient for that environmental factor and allow the organisms to choose where along the gradient they would prefer to be. If the organisms collect at certain positions along the gradient, we can see that the particular variable is significant to the organism.

In this exercise you will apply the scientific method. You will test the hypothesis that animals will respond to environmental gradients by congregating at specific positions along the gradient. You will set up an experimental situation and carefully and accurately collect data. You will then analyze the data you collect to determine which environmental variables are significant to the organism and where along the gradient they prefer to be.

During this exercise you will work in an assigned group. Each group works with one variable, such as light, pH, gravity, or temperature, and determines how the organism (brine shrimp) responds. Each group

1. places brine shrimp in the test apparatus (plastic tubing or trough).
2. adjusts the apparatus to establish the specific environmental gradient assigned to it.
3. allows the brine shrimp sufficient time to move to their preferred position along the gradient.
4. collects data concerning population density at five positions along the gradient.
5. reports its data to the class.
6. records data collected by other groups.
7. interprets all the data reported.

Experimental Design and Data Collection

Although it may appear simple to count the organisms present at each point along the gradient, there are several problems that may cause inaccurate counts. Also the number of organisms will be very large. Therefore, it may be desirable to count a random sample of the organisms from each of the five positions along the gradient. Therefore, your instructor will discuss possible ways to collect the samples and count the number of organisms in each sample. The following items need to be considered:

1. How should the organisms be removed from the apparatus so that one sample is totally isolated from others?
2. Should you count every individual, or should you sample your populations from each section?
3. How do you make sure that you are counting only living organisms?

Procedure—Setting Up the Experiment

1. The class will be divided into five groups: group 1 (control), group 2 (pH), group 3 (temperature), group 4 (light), and group 5 (gravity).
2. Each group needs to obtain its test apparatus which will be either a specific piece of tubing or a plastic trough.
3. Fill the test apparatus with brine shrimp from a well-mixed culture.
4. Adjust your specific variable as described below and allow the container to remain undisturbed for 30 minutes.

Specific Instructions for Each Group

Group 1 (Control)

The **control** group should have no gradients from one end to the other. Other groups will compare their data to yours to see if their data differ from yours. You want to make sure that all regions of the test apparatus have exactly the same conditions. The apparatus should be horizontal, have no access to light, have no difference in temperature along its length, and have no difference in pH from one end to the other. Leave the container undisturbed for 30 minutes, then collect your data by the method agreed to at the beginning of the lab.

Group 2 (pH)

You are working with one variable, pH. You want to establish a pH gradient along the length of your test apparatus. Your apparatus should be horizontal, not have access to light, and be the same temperature from one end to the other. To establish the pH gradient, use a hypodermic syringe to slowly inject 0.5 ml of 1% HCl into one end of the container. Next, use a different syringe to inject 1 ml of 1% KOH into the other end. *Be careful when removing the needle from the tube. A small amount of liquid may spray out.* Leave the container undisturbed for 30 minutes, then collect your data by the method agreed to at the beginning of the lab. You will also need to determine the pH of each of the 5 samples you collect.

Group 3 (Temperature)

You are working with one variable, temperature. You want to establish a temperature gradient along the length of your test apparatus. Your apparatus should be horizontal, not have access to light, and be the same pH from one end to the other. To establish the temperature gradient, cover the left end of the container with a plastic bag of crushed ice and place an infrared heat lamp 30 cm above the other end. Leave the container undisturbed for 30 minutes, then collect your data by the method agreed to at the beginning of the lab. You will also need to record the temperature of each of the five samples you collect.

Group 4 (Light)

You are working with one variable, light. You want to establish a light gradient along the length of your test apparatus. Your apparatus should be horizontal and have the same temperature and pH from one end to the other. To establish the light gradient, place a source of light at one end of the apparatus. The apparatus will need to be shielded from the lights from the room so that the only source of light available to the brine shrimp is coming from one end of the apparatus. The source of light should either be a fluorescent lamp or be placed far enough away from the apparatus so that it does not heat up one end of the apparatus and accidently set up a temperature gradient. Leave the container undisturbed for 30 minutes, then collect your data by the method agreed to at the beginning of the lab. You will also need to record the distance each of the five samples is from the source of light.

Group 5 (Gravity)

You are working with one variable, gravity. You want to establish a gravity gradient along the length of your test apparatus. Your apparatus should not have access to light and should be the same pH and temperature from one end to the other. To establish the gravity gradient, position the apparatus so that one end is much higher than the other. Leave the apparatus undisturbed for 30 minutes, then collect your data by the method agreed to at the beginning of the lab. You will also need to record the elevation of each of the five samples you collect.

Data Gathering

After allowing 30 minutes for your brine shrimp to respond to the environmental gradient you established, do the following:

1. Divide the apparatus into five equal sections so that organisms are unable to swim from one section to the next. Section your apparatus as follows.

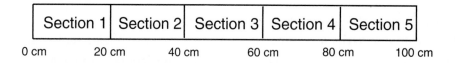

2. Empty the contents of each section into a separate beaker.
3. Label your beakers so that you can identify which beaker came from each section of the apparatus.
4. Record the pH and temperature of each beaker.
5. Measure the volume of water in each of the beakers.
6. Count the number of individuals in each sample by the method agreed to at the beginning of the lab.
7. Divide the number of brine shrimp in your sample by the number of milliliters of water in that portion of the apparatus. This will give you the number of organisms per mL of water.

$$\text{Brine shrimp per mL} = \frac{\text{number of brine shrimp counted in the section}}{\text{number of mL of water in the section}}$$

8. Report your data and record data from all other groups in table 41.1.
9. Interpret the data collected. Do brine shrimp respond to light, temperature, pH, or gravity? How do you know? Describe the preferred habitat of brine shrimp.

Table 41.1　　Data Sheet

Team	Section of the Apparatus				
	I	II	III	IV	V
1. Control	**Left End** pH_____ Temp._____ mL H_2O_____ organisms counted_____ org/mL_____	pH_____ Temp._____ mL H_2O_____ organisms counted_____ org/mL_____	pH_____ Temp._____ mL H_2O_____ organisms counted_____ org/mL_____	pH_____ Temp._____ mL H_2O_____ organisms counted_____ org/mL_____	**Right End** pH_____ Temp._____ mL H_2O_____ organisms counted_____ org/mL_____
2. pH	**Acid** pH_____ Temp._____ mL H_2O_____ organisms counted_____ org/mL_____	pH_____ Temp._____ mL H_2O_____ organisms counted_____ org/mL_____	pH_____ Temp._____ mL H_2O_____ organisms counted_____ org/mL_____	pH_____ Temp._____ mL H_2O_____ organisms counted_____ org/mL_____	**Base** pH_____ Temp._____ mL H_2O_____ organisms counted_____ org/mL_____
3. Temperature	**Cold** pH_____ Temp._____ mL H_2O_____ organisms counted_____ org/mL_____	pH_____ Temp._____ mL H_2O_____ organisms counted_____ org/mL_____	pH_____ Temp._____ mL H_2O_____ organisms counted_____ org/mL_____	pH_____ Temp._____ mL H_2O_____ organisms counted_____ org/mL_____	**Hot** pH_____ Temp._____ mL H_2O_____ organisms counted_____ org/mL_____
4. Light	**Dark** pH_____ Temp._____ mL H_2O_____ organisms counted_____ org/mL_____	pH_____ Temp._____ mL H_2O_____ organisms counted_____ org/mL_____	pH_____ Temp._____ mL H_2O_____ organisms counted_____ org/mL_____	pH_____ Temp._____ mL H_2O_____ organisms counted_____ org/mL_____	**Light** pH_____ Temp._____ mL H_2O_____ organisms counted_____ org/mL_____
5. Gravity	**Bottom** pH_____ Temp._____ mL H_2O_____ organisms counted_____ org/mL_____	pH_____ Temp._____ mL H_2O_____ organisms counted_____ org/mL_____	pH_____ Temp._____ mL H_2O_____ organisms counted_____ org/mL_____	pH_____ Temp._____ mL H_2O_____ organisms counted_____ org/mL_____	**Top** pH_____ Temp._____ mL H_2O_____ organisms counted_____ org/mL_____

Calculate the number of organisms per milliliter as follows:

$$\text{Organisms / mL} = \frac{\text{organisms counted in the sample}}{\text{total milliliters in the section of the apparatus}}.$$

The Effect of Abiotic Factors on Habitat Preference

Name_____ Lab section _____

Your instructor may collect these end-of-exercise questions. If so, please fill in your name and lab section.

End-of-Exercise Questions

1. What is the purpose of the control?

2. Why is it necessary to have large numbers of organisms in your sample?

3. Were the brine shrimp equally distributed in the five sections of control at the completion of the experiment? Should they have been the same? Explain why or why not.

4. Use the chi-square test to determine if the number of organisms in the five sections of the control are significantly different. Use the same test of statistical significance to evaluate the other four sets of data.

	O Observed	E Expected	(O - E)	$(O - E)^2$	$\dfrac{(O - E)^2}{E}$
Control					
pH					
Temperature					
Light					
Gravity					

Chi-square value = _____

Degrees of freedom = _____

Probability = _____

5. How would you modify your procedures if you were to repeat this exercise?

6. Using your data, describe how you think brine shrimp respond to

pH. _____

temperature. _____

light. _____

gravity. _____

Experiment 42: Natural Selection

Invitation to Inquiry

Our species has been around for over 100,000 years. During this period our species has been subjected to natural selection. How might selecting agents be different today from what they were 100,000 years ago? Speculate on how the following things have affected the evolution of humans.

glasses and contact lenses

antibiotics to control bacterial disease

availability of food

Introduction

The success of an individual within a species is determined by a variety of conditions and events in the life of the organism. The characteristics an individual has may determine whether it will survive and reproduce. The phenotype of an organism consists of the physical, behavioral, and physiological characteristics displayed by the organism. Most of these characteristics are determined by genes which can be passed from one generation to the next. From an evolutionary perspective the most successful individuals are those that reproduce and pass the largest number of copies of their genes on to the next generation. Those that pass on many copies of their genes are selected for and those that do not are selected against. The individual environmental factors that determine if an individual survives and reproduces are known as **selecting agents**. **Natural selection** is a term used to describe any natural events that determine which individuals within a species have the opportunity to pass their genes to the next generation. The result of natural selection is a change in the frequency of certain genes found in the species. **Evolution** is the continuous genetic adaptation of a population of organisms to its environment. Therefore, natural selection results in evolution.

In this exercise we will look at three different mechanisms by which natural selection can influence which individuals pass their genes to the next generation: differential survival, differential reproductive rates, and differential mate selection.

Differential Survival
Background Information

Several thousand years ago the survival of any individual human depended on an ability to locate food and avoid predators. Food was often in short supply, and individuals were forced to experiment. They would try new kinds of food if they were desperate. Many kinds of plants produce chemicals that taste bad and are toxic to the organisms that eat them. Humans were also preyed upon by other kinds of animals. Good eyesight was useful for avoiding predators and locating food. The ability to obtain adequate food was important for survival and reproduction because poorly nourished individuals had a much lower chance of successfully reproducing.

In the following activities we will simulate how the ability to taste, see, and obtain food could influence the number of offspring a particular individual human could produce.

371

Procedure

Avoiding Toxic Plants

All individuals will be issued a piece of PTC paper to taste.

1. Place the PTC paper in your mouth. You are either a taster or a nontaster. It will be easy for you to determine this.
   a. Tasters will be able to identify foods that contain certain common kinds of plant toxins. Therefore they will have a greater number of children that survive because they will not feed toxic plants to their children. During their lifetime each taster will have 10 children that will live.
   b. Nontasters will not be able to identify the toxins, will feed toxic plants to their children, and many of their offspring will die. Thus they will have only five offspring that live.
2. Record the number of offspring you will have in the space provided on the data sheet on page 375.

Spotting Danger and Locating Food

Because eyesight is important in spotting danger and locating food, the number of offspring produced will be related to how well people can see.

1. To simulate this situation the instructor will hold up a card with a word written on it. Write the word on the data sheet on page 375.
   a. Those who correctly identify the word will have 10 offspring.
   b. Those who get part of the word correct will have five offspring.
   c. Those who cannot identify the word will have zero offspring.
2. Record the number of offspring you will have on the data sheet on page 375.

Health and Nutrition

Nutritional status is important in determining how many offspring an individual is likely to have survive.

1. To simulate nutritional status write down what you ate for breakfast this morning in the space provided on the data sheet on page 375.
   a. Individuals who had breakfasts that included at least three of the major kinds of food items—(1)cereals, (2)milk products, (3)fruit, (4)vegetables, or (5)meat—will have 10 children.
   b. Individuals who ate breakfast but had only one or two of the major food groups will have five children.
   c. Individuals who had no breakfast (coffee, soft drinks, etc., do not qualify as breakfast) will have no offspring.
2. Record the number of offspring you have on the data sheet on page 375 and complete the analysis activities on the data sheet on page 376.

Differential Reproductive Rates

Background Information

The "fitness" of an individual organism is determined by the number of offspring the individual is able to produce. Each offspring produced by sexual reproduction is carrying half its genes from each parent. Therefore, each offspring produced by a parent allows the parent to pass its genes on to the next generation. Those individuals that pass more copies of their genes to the next generation are being selected for. The phenotype of an organism consists of the characteristics that can be observed. The phenotype is determined in part by the genes an organism has and in part by the environment of the organism.

Procedure

1. In this exercise you will be given a playing card that will represent your phenotype. The playing card represents an important feature of the organism that will determine the likelihood that the individual will reproduce.
2. Males and females in the class will have an opportunity to select mates and produce offspring. The object of the game is to produce the largest number of offspring in 10 generations.
3. No individual may have more than one mate. Unmated individuals will have no offspring. Any disputes about access to potential mates will be settled by the instructor based on the phenotype (playing card) of the disputing persons.
4. Move around the room and choose your mate based on your phenotype (playing card) and the phenotype of the potential mate. At this point you don't know what the best combinations are, but some combinations of phenotypes will produce many offspring and others will produce none.
5. Once all individuals have had an opportunity to choose a mate, the instructor will tell each couple how many offspring they produced. The instructor will use a consistent method of determining the number of offspring based on the combination of cards held by the couple. The combination of cards will result in three, two, one, or zero offspring.
6. At the end of each round, record the number of offspring you produced on the data sheet on page 375.
7. At the end of each round, all individuals may stay with the same mate or choose new mates. Remember your goal is to produce as many offspring as possible.
8. Continue this process through 10 rounds.
9. After the tenth round the instructor will explain the rules if they have not already been figured out.
10. Complete the analysis activities on the data sheet on page 376.

Differential Mate Selection—Lek Mating Systems

Background Information

In Lek Mating systems males stake out territories usually in the presence of other males. The females evaluate the characteristics of the various males and choose which male they will mate with from among all the males present. Usually only a small number of the males are chosen, and the other males do not mate. Males can have several mates during any mating season. The males that are chosen are selected for and have a greater chance of passing their genes on to the next generation.

Procedure

1. In a lek mating system, because females decide which males show the most desirable characteristics, the females will have special rules that will be involved in deciding which males will be chosen for mating and how many offspring they will have.
   a. The instructor will provide the females in the class with special rules that will determine which males are most suitable for mates.
   b. Each male will be issued a meterstick.
   c. Each person will be issued a playing card to be used in case disputes arise.
2. Males will distribute themselves around the room but must be at least 2 meters apart.
3. Females will choose mates by standing near them.
4. Any disputes about where a male may stand or which females have access to a specific male will be decided by the instructor based on the playing cards held by the individuals who are in dispute.
5. Each female will produce one offspring per year. Each male can produce as many offspring as he has females, but the maximum number of females per male is five. Each individual will record the number of offspring they will have per year.
6. At the end of each year the males may redistribute themselves and females must choose again.
7. Repeat for 10 rounds.
8. After the tenth round the instructor will explain the rules by which the females choose males.
9. Complete the analysis activities on the data sheet on page 377.

Natural Selection

Name _____ Lab section_____

Your instructor may collect these end-of-exercise questions. If so, please fill in your name and lab section.

Data Sheet and End-of-Exercise Questions

Natural Selection Scorecard

Differential Survival	
Offspring Produced	
Avoiding toxic plants_____ Spotting danger_____ Health and nutrition _____ Total _____	Write the word you saw here. Write the breakfast you ate here.

Differential Reproductive Rates	Differential Mate Selection
Offspring produced	Offspring produced
Year 1	Year 1
Year 2	Year 2
Year 3	Year 3
Year 4	Year 4
Year 5	Year 5
Year 6	Year 6
Year 7	Year 7
Year 8	Year 8
Year 9	Year 9
Year 10	Year 10
Total	Total

Differential Survival Analysis

The maximum number of offspring possible from this series of simulations is 30. Complete the following chart.

Number of Offspring	Number of People in Class That Had This Number of Offspring
30	
25	
20	
15	
10	
5	
0	

1. Which individuals in the class had genes that allowed them to be selected for?_____
   Which individuals were selected against? _____
2. Could individuals do anything to improve their chances of reproducting? _____

Differential Reproductive Rates Analysis

The maximum number of offspring possible is 30. Complete the following chart.

Number of Offspring	Number of People in Class That Had This Number of Offspring
28-30	
25-27	
22-24	
19-21	
16-18	
13-15	
10-12	
7-9	
4-6	
1-3	
0	

3. Based on the cards they held did all of the people in the class have the same opportunity to reproduce? _____

Differential Mate Selection Analysis

The maximum number of offspring possible is 50. Complete the following chart.

Number of Offspring	Number of People in Class That Had This Number of Offspring
46-50	
41-45	
36-40	
31-35	
26-30	
21-25	
16-20	
11-15	
6-10	
1-5	
0	

4. Were some males more successful than others?_____
5. How much of their success was determined by genes?_____

6. Describe two human characteristics presumed to be determined by genes that would lower a person's reproductive success.

7. Many eyesight characteristics are inherited (color blindness, astigmatism, near sightedness). Compared to 1000 years ago, do you feel these genes are being selected against more or less strongly? Explain your answer.

8. In many studies observers consider human individuals with symmetrical facial features to be more beautiful than those who have some degree of asymmetry. How might facial symmetry or lack of symmetry affect a person's reproductive success?

9. If an organism's reproductive fitness is determined by the genes it inherited, can all individuals have an equal chance of reproducing? Explain you answer.

10. If in a lek mating system the genes that determined the behavior of the females mutated so that they behaved differently, would the same males be successful? Explain your answer.

Experiment 43: Roll Call of the Animals

Invitation to Inquiry

We often overlook the many kinds of organisms that share our homes with us. Look in the light fixtures of your home or apartment and collect all the organisms you find there. Arrange them on a sheet of white paper. Arrange them into logical categories. You do not need to be able to tell exactly what they are. Just use obvious characteristics to sort them. How many different kinds of animals did you find?

Background

There is great diversity within the animal kingdom. No one can be expected to recognize all of the different organisms. You already know the differences between cows and dogs, birds and people, and snakes and frogs. You place these animals in categories based on certain differences in the characteristics that each possesses. If you know what critical traits to look for, it is possible to separate any animal into its proper taxonomic category. A taxonomic category is a group of closely related organisms that have evolved along similar lines. Recognizing the important characteristics that differentiate organisms into natural categories is the basis of the science of **taxonomy**, the classifying and naming of organisms.

The ranking order of classification groups (*taxons*) from the most inclusive through the least inclusive is as follows.

> Domain
> Kingdom
> Phylum
> Class
> Order
> Family
> Genus
> Species

Organisms in the same kingdom are very broadly similar; those in the same phylum are more similar to each other than those in other phyla. Those organisms in the same genus are quite similar, more closely so than those in another genus of the same family.

Each species of organism has a **scientific name** that is often descriptive and employ two terms: the genus name followed by a specific epithet. Since two names are used to identify a species, it is called a binomial nomenclature. The genus name is always capitalized, and the specific epithet is never capitalized, and both the genus and the specific epithet are underlined or set in italics. For example, the leopard frog is scientifically identified as *Rana pipiens*. You belong to the species *Homo sapiens*.

You can become acquainted with most of the important groups of the animal kingdom through this laboratory experience. By using a dichotomous key you will become familiar with some of the important characteristics used to classify animals. Although it is impossible to demonstrate all of the kinds of animals, you will have an opportunity to see examples of the major phyla and classes of animals. During this lab exercise you will

       1. use the dichotomous key provided to identify the various specimens.

       2. record the phylum or class to which each animal belongs.

Procedure

A **dichotomous key** is a tool used to help determine the taxonomic category to which a specific organism belongs. It consists of a series of pairs of statements that require you to place the organism into one of two categories. Each choice will lead you to another pair of statements until you have identified the animal's taxonomic category. For each of the organisms on display in the lab begin at step 1 of the key and proceed through the key until you have identified the specimen. When you have determined the name of the organism, write the underlined name (from the key) opposite the appropriate number on the answer sheet provided at the end of this exercise.

You may begin your work at any station and proceed to any available station thereafter; just be careful to place your answer at the appropriate number on your answer sheet.

Dichotomous Key

1. a. Irregular-shaped body; structure with many pores—<u>Phylum Porifera</u> (e.g., sponge)

   b. Regular-shaped body (with right and left halves or a cylindrical shape ------------------------2

2. a. **Radial symmetry** (disk-shaped or barrel-shaped)--4

   b. **Bilateral symmetry** (similar right and left body halves)--3

3. a. Animal has internal skeleton --19

   b. Animal has external skeleton or no apparent skeleton ---------------------------------- -------6

4. a. Body hard; arms extend from a central disc; or spines present—<u>Phylum Echinodermata</u> -----18

   b. Soft body; little or no color—<u>Phylum Coelenterata</u>--5

5. a. Saucer-shaped transparent body with small tentacles—<u>Class Scyphozoa</u> (e.g., jellyfish)

   b. Barrel-shaped body; tentacles at one end—<u>Class Anthozoa</u> (e.g., sea anemone)

6. a. Hard outer covering --10

   b. No hard outer covering ---7

7. a. Body flattened—<u>Phylum Platyhelminthes</u> --8

   b. Body not flattened ---9

8. a. Smooth, nonsegmented body—<u>Class Trematoda</u> (e.g., liver fluke)

   b. Apparently segmented, flattened body—<u>Class Cestoda</u> (e.g., tapeworm)

9. a. Nonsegmented --11

   b. **Segmented body**—<u>Phylum Annelida</u> (e.g., earthworm)

10. a. Body has jointed legs—<u>Phylum Arthropoda</u> --14

b. Body inside of shell is soft; has no jointed legs—<u>Phylum Mollusca</u> ---------------------------13

11. a. Tentacles or other appendages present --12

    b. Body long and tubular; no appendages—<u>Phylum Nematoda</u>

12. a. Appears as snail without shell—<u>Class Gastropoda</u> (e.g., slug)

    b. Tentacles and eyes present—<u>Class Cephalopoda</u> (e.g., squid, octopus)

13. a. Bivalved shell (two halves)—<u>Class Bivalvia</u> (e.g., clam)

    b. Univalved shell (single unit)—<u>Class Gastropoda</u> (e.g., snails)

14. a. Jointed appendages on most body sections --15

    b. Jointed appendages on certain body segments; not all appendages are legs--------------------16

15. a. One pair of legs per body segment—<u>Class Chilopoda</u> (e.g., centipede)

    b. Two pairs of legs per body segment—<u>Class Diplopoda</u> (e.g., millipede)

16. a. Two pairs of antennae; large claws often present—<u>Class Crustacea</u> (e.g., crab)

    b. One pair of antennae or none; no large claws---17

17. a. Four pairs of legs; no antennae or wings—<u>Class Arachnida</u> (e.g., spider)

    b. Three pairs of legs; wings present—<u>Class Insecta</u> (e.g., insects)

18. a. Arms present; body surface knobby—<u>Class Asteroidea</u> (e.g., sea stars)

    b. Many-spined animal; resembles a pincushion—<u>Class Echinoidea</u> (e.g., sea urchin)

19. a. Fish-like, flattened body; appendages fin-like not jointed--20

    b. Not fish-like; body not flattened; appendages jointed or absent----------------------------------21

20. a. Scales on body do not overlap; skeleton of cartilage—<u>Class Chondrichthyes</u> (sharks, stingray)

    b. Scales on body overlap; skeleton bony—<u>Class Osteichthyes</u> (e.g., bony fishes)

21. a. Body covered by scales; zero or four legs—<u>Class Reptilia</u> (e.g., snake, lizard, turtle)

    b. Body not covered by scales --22

22. a. Claws absent—<u>Class Amphibia</u> (e.g., Frogs, toads, and salamanders)

    b. Claws or nails present on toes; skin covered with feathers or hair ----------------------------23

23. a. Feathered; claws present—<u>Class Aves</u> (e.g., birds)

    b. Hair present—<u>Class Mammalia</u> (e.g., mammals)

Results

1. _____

2. _____

3. _____

4. _____

5. _____

6. _____

7. _____

8. _____

9. _____

10. _____

11. _____

12. _____

13. _____

14. _____

15. _____

16. _____

17. _____

18. _____

19. _____

20. _____

21. _____

22. _____

23. _____

Experiment 44: Special Project

The *Special Project* (SP) is an independent investigation that is accomplished outside of the classroom and laboratory room with ordinary and everyday devices. *No laboratory equipment* will be available for this project since the SP is intended to be an unstructured "kitchen science" investigation. You will need to take this your-stuff-only restriction into consideration when deciding what you are going to investigate as you *must* use only available equipment to do the investigation. This restriction is an important part of the process. The SP is an opportunity to explore that area of science that interests you most and in a real world situation. You are free to pursue any science concept for your experiment, but your project should not simply mimic one of the regular experiments assigned in class or one from the laboratory manual. As best you are able to, you are to play the role of an original, creative thinker during this investigation.

Important: An SP proposal must be submitted to your laboratory instructor and *approved* before beginning any work. This is necessary for safety considerations, to avoid project duplications, and to ensure that your project satisfies the intent of the assignment. Please read the guidelines below and also read the evaluation sheet to make sure you understand the intent of the SP. You will be informed of the due date for the proposal, which should briefly include the following:

- Name(s) of the person(s) working as an individual or as a team.

- A brief statement of the question to be answered.

- What equipment will be used in what procedure to find the answer.

- How measurements will be made; how data will be collected.

You will have about 30 days to conduct the SP outside of class. One lab period has been established as a reporting session, a time when the experimental findings will be presented in both oral and written formats. The oral presentation will not be as detailed as the written one; however, it is the only way the rest of the class will be able to benefit from your efforts. It therefore warrants some thought and creativity.

The SP **oral presentation** should include the following as possible:

- Reporting responsibilities should be equally divided among all team members.

- Describe the question to be answered and how the investigation was conducted.

- The independent and dependent variables should be identified.

- Use display boards to show diagrams, data tables, and graphs.

- An interpretation of results using graphic analysis (mathematical model).

- Explanations for the deviation (if any) of results from what was expected.

- Concluding statements.

The SP **written report** should be typed following the outline below. An appropriate title for your project should be in the center of the first page with the names of each team member in the upper right hand corner. Each of the following sections should be included in your report prefaced with the appropriate heading.

Purpose: A brief statement of the question that was investigated.

Apparatus: A diagram of the equipment with all parts labeled showing the experimental setup.

Procedure: The sequence for conducting the experiment stated in brief sentences. The independent and dependent variables should also be clearly identified, including a short statement of how the independent variables were controlled.

Raw Data: The values measured directly from the experiment with data from as many trials as judged necessary (a minimum of three trials is required). Data should be organized into neat tables with the units (m, kg, s, etc.) clearly labeled. This section should be the original handwritten data sheet written at the time of the experiment.

Evaluation Of Data: Begin with a presentation of findings via graphs and formal data tables. Formal data tables include averaged values of your multiple trials as well as processed data in a spreadsheet format. Processed data refers to calculated values derived by inserting experimental values into various equations. State what equations were used and identify the symbols used in the equations. If repetitive calculations are performed, show only one example calculation. All other calculated values will appear in your formal data tables. Graphs should be labeled with the variables on the appropriate axes and units indicated clearly. Interpret your graphs with statements of relationships between the variables. These statements need to be complete English sentences. A mathematical model for the graph should be found (if possible) and stated in this section also. An equation for a line with the slope and *y*-intercept given in the proper units is an example of such a mathematical model. This section also should contain a statement describing the quality of the results.

Conclusion: Results are compared to what was expected and plausible explanations offered for any deviation. The meaning of the slope of a graph and any equations derived from graphical analysis are also stated here. You should also state whether the goals of your experiment were accomplished or not.

Note: *As the due date for the SP proposal and the presentation day approaches please feel free to contact your laboratory instructor for individual help, advice, or encouragement.*

SP Evaluation Scorecard

Team Members:

Proposal (5 points)--- _____
    Turned in on time and clearly stated.

Format of Report (4 points)--- _____
    1. Group names, title.
    2. Each section of report clearly labeled, neat, and organized.

Purpose of Investigation (2 points)-- _____
    Question to be answered by experiment is clearly identified and stated.

Procedure (6 points)--- _____
    1. Independent and dependent variables are clearly defined and controlled.
    2. Clear, brief sequence of steps.
    3. Diagram(s) drawn with all components labeled.

Raw Data (6 points)-- _____
    1. Measurement data organized into neat tables.
    2. Values are clearly labeled.
    3. Multiple trials.

Evaluation Of Data (12 points)-- _____
    1. Tables and/or sample calculations.
    2. Graphs; variables on appropriate axes, use of units.
    3. Interpretation of graphs
        a. Written statement of relationship.
        b. Mathematical model (equation, units on slope).

Conclusion (12 points)-- _____
    1. Written explanation (English sentences) of relationships.
    2. Meaning of slope in terms of experimental question.
    3. General equation included.
    4. Reasonable explanation for divergent results (when applicable).

Presentation (3 points)--- _____
    1. Display clear and understandable.
    2. Team functioned well together.
    3. Team seemed knowledgeable in their presentation.

Total Points--- _____

Appendix I: The Simple Line Graph

An equation describes a relationship between variables, and a graph helps you "picture" this relationship. A line graph pictures how changes in one variable go with changes in a second variable, that is, how the two variables change together. One variable usually can be easily manipulated; the other variable is caused to change in value by manipulation of the first variable. The *manipulated* variable is known by various names (*independent, input, or cause variable*), and the *responding* variable is known by various related names (*dependent, output, or effect variable*). The manipulated variable is usually placed on the horizontal or *x*-axis of the graph, so you can also identify it as the *x-variable*. The responding variable is placed on the vertical or *y*-axis. This variable is identified as the *y-variable*.

The graph in Appendix figure I.1 shows the mass of different volumes of water at room temperature. Volume is placed on the *x*-axis because the volume of water is easily manipulated and the mass values change as a consequence of changing the values of volume. Note that both variables are named, and the measuring unit for each variable is identified on the graph.

The graph also shows a number scale on each axis that represents changes in the values of each variable. The scales are usually, but not always, linear. A *linear* scale has equal intervals that represent equal increases in the value of the variable. Thus, a certain distance on the *x*-axis to the right represents a certain increase in the value of the *x*-variable. Likewise, certain distances up the *y*-axis represent certain increases in the value of the *y*-variable. In the example, each mark has a value

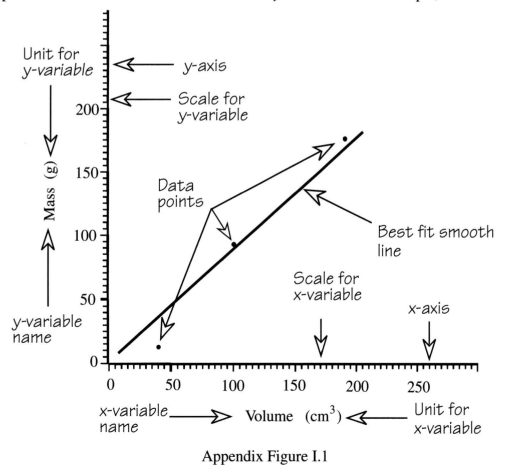

Appendix Figure I.1

of five. Scales are usually chosen in such a way that the graph is large and easy to read. The *origin* is the only point where both the *x*- and *y*-variables have a value of zero at the same time.

The example graph has three data points. A *data point* represents measurements of two related variables that were made at the same time. For example, a volume of 190 cm^3 of water was found to have a mass of 175 g. Locate 190 cm^3 on the *x*-axis and imagine a line moving straight up from this point on the scale (each mark on the scale has a value of 5 cm^3). Now locate 175 g on the *y*-axis and imagine a line moving straight out from this point on the scale (again, note that each mark on this scale has a value of 5 g). Where the lines meet is the data point for the 190 cm^3 and175 g measurements. A data point is usually indicated with a small dot or an x; a dot is used in the example graph.

A "best-fit" smooth, straight line is drawn as close to all the data points as possible. If it is not possible to draw the straight line *through* all the data points (and it usually never is), then a straight line should be drawn that has the same number of data points on both sides of the line. Such a line will represent a best approximation of the relationship between the two variables. The *origin* is also used as a data point in the example because a volume of zero will have a mass of zero. In any case, the dots are *never* connected as in dot-to-dot sketches. For most of the experiments in this lab manual a set of perfect, error-free data would produce a straight line. In such experiments it is not a straight line because of experimental error, and you are trying to eliminate the error by approximating what the relationship should be.

The smooth, straight line tells you how the two variables get larger together. If the scales on both the axes are the same, a 45° line means that the two variables are increasing in an exact direct proportion. A more flat or more upright line means that one variable is increasing faster than the other. The more you work with graphs, the easier it will become for you to analyze what the slope means.

Appendix II: The Slope of a Straight Line

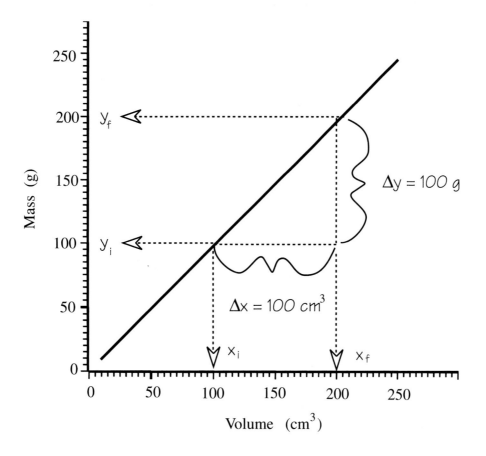

Appendix Figure II.1

One way to determine the relationship between two variables that are graphed with a straight line is to calculate the **slope.** The slope is a ratio between the changes in one variable and the changes in the other. The ratio is between the changes in the value of the x-variable compared to the changes in the value of the y-variable. The symbol Δ (Greek letter Delta) means "change in," so the symbol Δx means "change in x." The first step in calculating the slope is to find out how much the x-variable is changing (Δx) in relation to how much the y-variable is changing (Δy). You can find this relationship by first drawing a dashed line to the right of the straight line so that the x-variable has increased by some convenient unit as shown in the example in Appendix figure II.1. *Where you start or end this dashed line will not matter since the ratio between the variables will be the same everywhere on the graph line.* However, it is very important to remember when finding a slope of a graph to *avoid using data points* in your calculations. Two points whose coordinates are easy to find should be used instead of data points. One of the main reasons for plotting a graph and drawing a best-fit straight line is to smooth out any measurement errors made. Using data points directly in calculations defeats this purpose.

The Δx is determined by subtracting the final value of the x-variable on the dashed line (x_f) from the initial value of the x-variable on the dashed line (x_i), or $\Delta x = x_f - x_i$. In the example graph above, the dashed line has a x_f of 200 cm³ and a x_i of 100 cm³, so Δx is 200 cm³ – 100 cm³, or 100 cm³. *Note that Δx has both a number value and a unit.*

389

Now you need to find Δy. The example graph shows a dashed line drawn back up to the graph line from the x-variable dashed line. The value of Δy is $y_f - y_i$. In the example, $\Delta y = 200$ g – 100 g. The slope of a straight graph line is the ratio of Δy to Δx, or

$$\text{Slope} = \frac{\Delta y}{\Delta x}.$$

In the example,

$$\text{Slope} = \frac{100 \text{ g}}{100 \text{ cm}^3}$$

or

$$\text{Slope} = 1 \text{ g}/\text{cm}^3.$$

Thus the slope is 1 g/cm^3, and this tells you how the variables change together. Since g/cm^3 is also the definition of density, you have just calculated the density of water from a graph.

Note that the slope can be calculated only for two variables that are increasing together (variables that are in direct proportion and have a line that moves upward and to the right). If variables change in any other way, mathematical operations must be performed to *change the variables into this relationship*. Examples of such necessary changes include taking the inverse of one variable, squaring one variable, taking the inverse square, and so forth.

Appendix III: Experimental Error

All measurements are subject to some uncertainty, as a wide range of errors can and do happen. Measurements should be made with great accuracy and with careful thought about what you are doing to reduce the possibility of error. Here is a list of some of the possible sources of error to consider and avoid.

Improper Measurement Technique. Always use the smallest division or marking on the scale of the measuring instrument, then estimate the next interval between the shown markings. For example, the instrument illustrated in Appendix figure III.1 shows a measurement of 2.45 units, and the .05 is estimated because the reading is about halfway between the marked divisions of 2.4 and 2.5. If you do not estimate the next smallest division you are losing information that may be important to the experiment you are conducting.

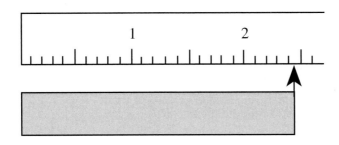

Appendix Figure III.1

Incorrect Reading. This is an error in reading (misreading) an instrument scale. Some graduated cylinders, for example, are calibrated with marks that represent 2.0 mL intervals. Believing that the marks represent 1.0 mL intervals will result in an incorrect reading. This category of errors also includes the misreading of a scale that often occurs when you are not paying sufficient attention to what you are doing.

Incorrect Recording. A personal mistake that occurs when the data are incorrectly recorded; for example, making a reading of 2.54 units and then recording a measurement of 2.45 units.

Assumptions About Variables. A personal mistake that occurs when there is a lack of clear, careful thinking about what you are doing. Examples are an assumption that water always boils at a temperature of 212° F (100° C), or assuming that the temperature of a container of tap water is the same now as it was 15 minutes ago.

Not Controlling Variables. This category of errors is closely related to the assumptions category but in this case means failing to recognize the influence of some variable on the outcome of an experiment. An example is the failure to recognize the role that air resistance might have in influencing the length of time that an object falls through the air.

Math Errors. This is a personal error that happens to everyone but penalizes only those who do not check their work and think about the results and what they mean. Math errors include not using significant figures for measurement calculations.

Accidental Blunders. Just like math errors, accidents do happen. However, the blunder can come from a poor attitude or frame of mind about the quality of work being done. In the laboratory, an example of a lack of quality work would be spilling a few drops of water during an experiment with an "Oh well, it doesn't matter" response.

Instrument Calibration. Errors can result from an incorrectly calibrated instrument, but these errors can be avoided by a quality work habit of checking the calibration of an instrument against a known standard, then adjusting the instrument as necessary.

Inconsistency. Errors from inconsistency are again closely associated with a lack of quality work habits. Such errors could result from a personal bias; that is, trying to "fit" the data to an expected outcome or using a single measurement when a spread of values is possible.

Whatever the source of errors, it is important that you recognize the error, or errors, in an experiment and know the possible consequence and impact on the results. After all, how else will you know if two seemingly different values from the same experiment are acceptable as the "same" answer or which answer is correct? One way to express the impact of errors is to compare the results obtained from an experiment with the true or accepted value. Everyone knows that percent is a ratio that is calculated from

$$\frac{\text{Part}}{\text{Whole}} \times 100\% \text{ of whole } = \% \text{ of part.}$$

This percent relationship is the basic form used to calculate a percent error or a percent difference.

The **percent error** is calculated from the *absolute difference* between the experimental value and the accepted value (the part) divided by the accepted value (the whole). Absolute difference is designated by the use of two vertical lines around the difference, so

$$\% \text{ Error } = \frac{|\text{Experimental value} - \text{Accepted value}|}{\text{Accepted value}} \times 100\%.$$

Note that the absolute value for the part is obtained when the smaller value is subtracted from the larger. For example, suppose you experimentally determine the frequency of a tuning fork to be 511 Hz, but the accepted value stamped on the fork is 522 Hz. Subtracting the smaller value from the larger, the percentage error is

$$\frac{|522 \text{ Hz} - 511 \text{ Hz}|}{522 \text{ Hz}} \times 100\% = 2.1\%.$$

You should strive for the lowest percentage error possible, but some experiments will have a higher level of percentage errors than other experiments, depending on the nature of the measurements required. In some experiments the acceptable percentage error might be 5%, but other experiments could require a percentage error of no more than 2%.

A true, or accepted, value is not always known, so it is sometimes impossible to calculate an actual error. However, it is possible in these situations to express the error in a measured quantity as a percent of the quantity itself. This is called a **percent difference**, or a percent deviation from the mean. This method is used to compare the accuracy of two or more measurements by seeing how consistent they are with each other. The percent difference is calculated from the *absolute difference* between one measurement and a second measurement, divided by the average of the two measurements. As before, absolute difference is designated by the use of two vertical lines around the difference, and

$$\% \text{ Difference} = \frac{|\text{One value} - \text{Another value}|}{\text{Average of the two values}} \times 100\%.$$

Appendix IV: Significant Figures

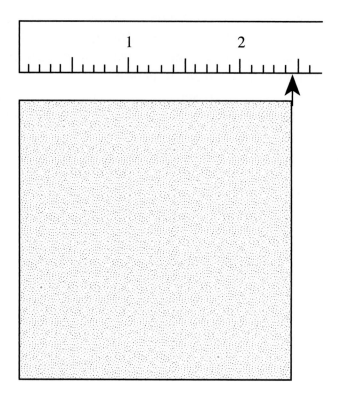

The numerical value of any measurement will always contain some uncertainty. Suppose, for example, that you are measuring one side of a square piece of paper as shown above. You could say that the paper is *about* 2.5 cm wide, and you would be correct. This measurement, however, would be unsatisfactory for many purposes. It does not approach the true value of the length and contains too much uncertainty. It seems clear that the paper width is larger than 2.4 cm but shorter than 2.5 cm. But how much larger than 2.4 cm? You cannot be certain if the paper is 2.44, 2.45, or 2.46 cm wide. As your best estimate, you might say that the paper is 2.45 cm wide. Everyone would agree that you can be certain about the first two numbers (2.4), and they should be recorded. The last number (0.05) has been estimated and is not certain. The two certain numbers, together with one uncertain number, represent the greatest accuracy possible with the ruler being used. The paper is said to be 2.45 cm wide.

A **significant figure** is a number that is believed to be correct with some uncertainty only in the last digit. The value of the width of the paper (2.45 cm) represents three significant figures. As you can see, the number of significant figures can be determined by the degree of accuracy of the measuring instrument being used. But suppose you need to calculate the area of the paper. You would multiply 2.45 cm $\times$ 2.45 cm, and the product for the area would be 6.0025 cm^2. This is a greater accuracy than you were able to obtain with your measuring instrument. *The result of a calculation can be no more accurate than the values being treated.* Because the measurement had only three significant figures (two certain, one uncertain), then the answer can have only three significant figures. Thus the area is correctly expressed as 6.00 cm^2.

There are a few simple rules that will help you determine how many significant figures are contained in a reported measurement.

Rule 1. All digits reported as a direct result of a measurement are significant.

Rule 2. Zero is significant when it occurs between nonzero digits. For example, 607 has three significant figures, and the zero is one of the significant figures.

Rule 3. In figures reported as *larger than the digit one,* the digit zero is *not significant* when it follows a nonzero digit to indicate place. For example, in a report that "23,000 people attended the rock concert," the digits 2 and 3 are significant, but the zeros are not significant. In this situation the 23 is the measured part of the figure, and the three zeros tell you an estimate of how many attended the concert, that is, 23 thousand. If the figure is a measurement rather than an estimate, then it is written *with a decimal point after the last zero* to indicate that the zeros *are* significant. Thus 23,000 has *two* significant figures (2 and 3), but 23,000. has *five* significant figures. The figure 23,000 means "about 23 thousand" but 23,000. means 23,000. and not 22,999 or 23,001. One way to show the number of significant figures is to use scientific notation, for example, 2.3×10^3 has two significant figures, 2.30×10^3 has three, and 2.300×10^4 has four significant figures. Another way to show the number of significant figures is to put a bar over the top of a significant zero if it could be mistaken for a placeholder.

Rule 4. In figures reported as *smaller than the digit one*, zeros after a decimal point that come before nonzero digits *are not* significant and serve only as place holders. For example, 0.0023 has two significant figures, 2 and 3. Zeros alone after a decimal point or zeros after a nonzero digit indicate a measurement, however, so these zeros *are* significant. The figure 0.00230, for example, has three significant figures since the 230 means 230 and not 229 or 231. Likewise, the figure 3.000 cm has four significant figures because the presence of the three zeros means that the measurement was actually 3.000 and not 2.999 or 3.001.

Multiplication and Division

When multiplying or dividing measurement figures, the answer may have no more significant figures than the *least* number of significant figures in the figures being multiplied or divided. This simply means that an answer can be no more accurate than the least accurate measurement entering into the calculation, and that you cannot improve the accuracy of a measurement by doing a calculation. For example, in multiplying 54.2 mi/hr $\times$ 4.0 hours to find out the total distance traveled, the first figure (54.2) has three significant figures but the second (4.0) has only two significant figures. The answer can contain only two significant figures since this is the weakest number of those involved in the calculation. The correct answer is therefore 220 miles, not 216.8 miles. This may seem strange since multiplying the two numbers together gives the answer of 216.8 miles. This answer, however, means a greater accuracy than is possible and the accuracy cannot be improved over the weakest number involved in the calculation. Since the weakest number (4.0) has only two significant figures the answer must also have only two significant figures, which is 220 miles.

The result of a calculation is **rounded** to have the same least number of significant figures as the least number of a measurement involved in the calculation. When rounding numbers, the last significant figure is increased by one if the number after it is five or larger. If the number after the last significant figure is four or less, the nonsignificant figures are simply dropped. Thus, if two significant figures are called for in the answer of the above example, 216.8 is rounded up to 220 because the last number after the two significant figures is 6, a number larger than 5. If the calculation result had been 214.8, the rounded number would be 210 miles.

Note that *measurement figures* are the only figures involved in the number of significant figures in the answer. Numbers that are **counted or defined** are not included in the determination of significant figures in an answer. For example, when dividing by 2 to find an average property of two objects, the 2 is ignored when considering the number of significant figures. Defined numbers are defined exactly and are not used in significant figures. For example, that a diameter is 2 times the radius is not a measurement. In addition, 1 kilogram is *defined* to be exactly 1000 grams, so such a conversion is not a measurement.

Addition and Subtraction

Addition and subtraction operations involving measurements, as with multiplication and division, cannot result in an answer that implies greater accuracy than the measurements had before the calculation. Recall that the last digit in a measurement is considered to be uncertain because it is the result of an estimate. The answer to an addition or subtraction calculation can have this uncertain number *no farther from the decimal place than it was in the weakest number involved in the calculation*. Thus when 8.4 is added to 4.926, the weakest number is 8.4 and the uncertain number is .4, one place to the right of the decimal. The sum of 13.326 is therefore rounded to 13.3, reflecting the placement of this weakest doubtful figure.

Example Problem

In Appendix III, "Experimental Error," an example was given of an experimental result of 511 Hz and an accepted value of 522 Hz, resulting in a calculation of

$$\frac{|522 \text{ Hz} - 511 \text{ Hz}|}{522 \text{ Hz}} \times 100\% = 2.1\%.$$

Since 522 – 511 is 11, the least number of significant figures of measurements involved in this calculation is *two*. Note that the 100 does not enter into the determination since it is not a measurement number. The calculated result (from a calculator) is 2.1072797, which is rounded off to have only two significant figures, so the answer is recorded as 2.1%.

Appendix V: Conversion of Units

The measurement of most properties results in both a numerical value and a unit. The statement that a glass contains 50 cm³ of a liquid conveys two important concepts: the numerical value of 50 and the reference unit of cubic centimeters. Both the numerical value and the unit are necessary to communicate correctly the volume of the liquid.

When working with calculations involving measurement units, *both* the numerical value and the units are treated mathematically. As in other mathematical operations, there are general rules to follow.

Rule 1. Only properties with *like units* may be added or subtracted. It should be obvious that adding quantities such as 5 dollars and 10 dimes is meaningless. You must first convert to like units before adding or subtracting.

Rule 2. Like or unlike units may be multiplied or divided and treated in the same manner as numbers. You have used this rule when dealing with area (length × length = length², for example, or cm × cm = cm²) and when dealing with volume (length × length × length = length³, for example, or cm × cm × cm = cm³).

You can use the above two rules to create a **conversion ratio** that will help you change one unit to another. Suppose you need to convert 2.3 kilograms to grams. First, write the relationship between kilograms and grams:

$$1000 \text{ grams} = 1.000 \text{ kg.}$$

Next, divide both sides by what you wish to convert *from* (kilograms in this example):

$$\frac{1000 \text{ g}}{1.000 \text{ kg}} = \frac{1.000 \text{ kg}}{1.000 \text{ kg}}.$$

One kilogram divided by one kilogram equals 1, just as 10 divided by 10 equals 1. Therefore, the right side of the relationship becomes 1:

$$\frac{1000 \text{ g}}{1.000 \text{ kg}} = 1.$$

The 1 is usually understood — that is, not stated — and the operation is called *canceling*. Canceling leaves you with the fraction 1000 g/1.000 kg, which is a conversion ratio that can be used to convert from kg to g. You simply multiply the conversion ratio by the numerical value and unit you wish to convert:

$$2.3 \text{ kg} \times \frac{1000 \text{ g}}{1.000 \text{ kg}} = 2300 \text{ g.}$$

The kg units cancel. Showing the whole operation with units only, you can see how you end up with the correct unit of g:

$$\text{kg} \times \frac{\text{g}}{\text{kg}} = \frac{\text{kg} \cdot \text{g}}{\text{kg}} = \text{g.}$$

Since you did obtain the correct unit, you know that you used the correct conversion ratio. If you had blundered and used an inverted conversion ratio, you would obtain:

$$2.3 \times \frac{1.000 \text{ kg}}{1000 \text{ g}} = 23 \frac{\text{kg}^2}{\text{g}},$$

which yields the meaningless, incorrect units of kg^2/g. Carrying out the mathematical operations on the numbers and the units will always tell you if you used the correct conversion ratio or not.

Example Problem

A distance is reported as 100.0 km and you want to know how far this is in miles.

Solution

First, you need to obtain a conversion factor from a textbook or reference book, which usually groups similar conversion factors in a table. Such a table will show two conversion factors for kilometers and miles: (a) 1.000 km = 0.621 mi and (b) 1.000 mi = 1.609 km. You select the factor that is in the same form as your problem. For example, your problem is 100.0 km = ? mi. The conversion factor in this form is 1.000 km = 0.621 mi.

Second, you convert this conversion factor into a **conversion ratio** by dividing the factor by what you want to convert *from*.

Conversion factor: 1.000 km = 0.621 mi

Divide factor by
what you want $\dfrac{1.000 \text{ km}}{1.000 \text{ km}} = \dfrac{0.621 \text{ mi}}{1.000 \text{ km}}$
to convert from:

Resulting
conversion ratio: $\dfrac{0.621 \text{ mi}}{\text{km}}$

The conversion ratio is now multiplied by the numerical value and unit you wish to convert.

$$100.0 \text{ km} \times \frac{0.621 \text{ mi}}{\text{km}}$$

$$100.0 \times 0.621 \frac{\text{km} \cdot \text{mi}}{\text{km}}$$

$$62.1 \text{ mi}$$

Appendix VI: Scientific Notation

Most of the properties of things that you might measure in your everyday world can be expressed with a small range of numerical values together with some standard unit of measure. The range of numerical values for most everyday things can be dealt with by using units (1's), tens (10's), hundreds (100's), or perhaps thousands (1,000's). But the universe contains some objects of incredibly large size that require some very big numbers to describe. The sun, for example, has a mass of about 1,970,000,000,000,000,000,000,000,000,000 kg. On the other hand, very small numbers are needed to measure the size and parts of an atom. The radius of a hydrogen atom, for example, is about 0.00000000005 m. Such extremely large and small numbers are cumbersome and awkward since there are so many zeros to keep track of, even if you are successful in carefully counting all the zeros. A method does exist to deal with extremely large or small numbers in a more condensed form. The method is called **scientific notation**, but it is also sometimes called *powers of ten* or *exponential notation* since it is based on exponents of 10. Whatever it is called, the method is a compact way of dealing with numbers that not only helps you keep track of zeros but also provides a simplified way to make calculations as well.

In algebra you save a lot of time (as well as paper) by writing (a $\times$ a $\times$ a $\times$ a $\times$ a) as a^5. The small number written to the right and above a letter or number is a superscript called an **exponent**. The exponent means that the letter or number is to be multiplied by itself that many times. For example, a^5 means "a" multiplied by itself five times, or a $\times$ a $\times$ a $\times$ a $\times$ a. As you can see, it is much easier to write the exponential form of this operation than it is to write out the long form.

Scientific notation uses an exponent to indicate the power of the base 10. The exponent tells how many times the base, 10, is multiplied by itself. For example:

$$10,000. = 10^4$$

$$1,000. = 10^3$$

$$100. = 10^2$$

$$10. = 10^1$$

$$1. = 10^0$$

$$0.1 = 10^{-1}$$

$$0.01 = 10^{-2}$$

$$0.001 = 10^{-3}$$

$$0.0001 = 10^{-4}$$

This table could be extended indefinitely, but this somewhat shorter version will give you an idea of how the method works. The symbol 10^4 is read as "ten to the fourth power" and means $10 \times 10 \times 10 \times 10$. Ten times itself four times is 10,000, so 10^4 is the scientific notation for 10,000. It is also equal to the number of zeros between the 1 and the decimal point. That is, to write the longer form of 10^4 you simply write 1, then move the decimal point four places to the *right*; hence ten to the fourth power is 10,000.

The powers of ten table also shows that numbers smaller than one have negative exponents. A negative exponent means a reciprocal:

$$10^{-1} = \frac{1}{10} = 0.1$$

$$10^{-2} = \frac{1}{100} = 0.01$$

$$10^{-3} = \frac{1}{1000} = 0.001$$

To write the longer form of 10^{-4}, you simply write 1 then move the decimal point four places to the *left*; hence ten to the negative fourth power is 0.0001.

Scientific notation usually, but not always, is expressed as the product of two numbers: (1) a number between 1 and 10 that is called the **coefficient** and (2) a power of ten that is called the **exponential**. For example, the mass of the sun that was given in long form earlier is expressed in scientific notation as

$$1.97 \times 10^{30} \text{ kg,}$$

and the radius of a hydrogen atom is

$$5.0 \times 10^{-11} \text{ m.}$$

In these expressions, the coefficients are 1.97 and 5.0, and the power of ten notations are the exponentials. Note that in both of these examples, the exponential tells you where to place the decimal point if you wish to write the number all the way out in the long form. Sometimes scientific notation is written without a coefficient, showing only the exponential. In these cases the coefficient of 1.0 is understood; that is, not stated. If you try to enter a scientific notation in your calculator, however, you will need to enter the understood 1.0 or the calculator will not be able to function correctly. Note also that 1.97×10^{30} kg and the expressions 0.197×10^{31} kg and 19.7×10^{29} kg are all correct expressions of the mass of the sun. By convention, however, you will use the form that has one digit to the left of the decimal.

Example Problem

What is 26,000,000 in scientific notation?

Solution

Count how many times you must shift the decimal point until one digit remains to the left of the decimal point. For numbers larger than the digit 1, the number of shifts tells you how much the exponent is increased, so the answer is 2.6×10^7, which means the coefficient 2.6 is multiplied by 10 seven times.

Example

What is 0.000732 in scientific notation? (Answer: 7.32×10^{-4})

Multiplication and Division

It was stated earlier that scientific notation provides a compact way of dealing with very large or very small numbers but provides a simplified way to make calculations as well. There are a few mathematical rules that will describe how the use of scientific notation simplifies these calculations.

To *multiply* two scientific notation numbers, the coefficients are multiplied as usual, and the exponents are *added* algebraically. For example, to multiply (2×10^2) by (3×10^3), first separate the coefficients from the exponentials,

$$(2 \times 3) \times (10^2 \times 10^3),$$

then multiply the coefficients and add the exponents,

$$6 \times 10^{(2+3)} = 6 \times 10^5.$$

Adding the exponents is possible because $10^2 \times 10^3$ means the same thing as $(10 \times 10) \times (10 \times 10 \times 10)$, which equals $(100) \times (1,000)$, or 100,000, which is expressed as 10^5 in scientific notation. Note that two negative exponents add algebraically, for example, $10^{-2} \times 10^{-3} = 10^{[(-2) + (-3)]} = 10^{-5}$. A negative and a positive exponent also add algebraically, as in $10^5 \times 10^{-3} = 10^{[(+5) + (-3)]} = 10^2$.

If the result of a calculation involving two scientific notation numbers does not have the conventional one digit to the left of the decimal, move the decimal point so it does, changing the exponent according to which way and how much the decimal point is moved. Note that the exponent increases by one number for each decimal point moved to the left. Likewise, the exponent decreases by one number for each decimal point moved to the right. For example, $938. \times 10^3$ becomes 9.38×10^5 when the decimal point is moved two places to the left.

To *divide* two scientific notation numbers, the coefficients are divided as usual, and the exponents are *subtracted*. For example, to divide (6×10^6) by (3×10^2), first separate the coefficients from the exponentials,

$$\left(\frac{6}{3}\right) \times \left(\frac{10^6}{10^2}\right)$$

then divide the coefficients and subtract the exponents,

$$2 \times 10^{(6-2)} \quad = \quad 2 \times 10^4.$$

Note that when you subtract a negative exponent, for example, $10^{[(3)-(-2)]}$, you change the sign and add, $10^{(3+2)} = 10^5$.

Example Problem

Solve the following problem concerning scientific notation:

$$\frac{\left(2 \times 10^4\right) \times \left(8 \times 10^{-6}\right)}{8 \times 10^4}.$$

Solution

First, separate the coefficients from the exponentials,

$$\frac{2 \times 8}{8} \quad \times \quad \frac{10^4 \times 10^{-6}}{10^4},$$

then multiply and divide the coefficients and add and subtract the exponents as the problem requires,

$$2 \times 10^{\{[(4)+(-6)]-[4]\}}.$$

Solving these remaining operations gives 2×10^{-6}.

Notes

Notes

Notes